VERTEBRATE ZOOLOGY AND EVOLUTION

VERTEBRATE ZOOLOGY AND EVOLUTION

– by –

B.N. Yadav

Professor and Head (Retd.)

P.G. Department of Zoology

Magadh University, Bodh-Gaya

(Bihar)

&

D. Kumar

M.Sc. (Gold Medalist), Ph.D.

Lecturer in Zoology

Bokaro Mahila College

Bokaro Steel City (Bihar)

2018

Daya Publishing House®

A Division of

Astral International Pvt. Ltd.

New Delhi - 110 002

First Published, 2000
Reprinted, 2018

ISBN: 978-81-7035-227-3 (Hardbound)

Published by : **Daya Publishing House®**
A Division of
Astral International Pvt. Ltd.
– ISO 9001:2015 Certified Company –
4736/23, Ansari Road, Darya Ganj
New Delhi-110 002
Ph. 011-43549197, 23278134
E-mail: info@astralint.com
Website: www.astralint.com

Preface to First Edition

It was perceived that there was scarcity of a good book on *"Vertebrate Zoology and Evolution"* for the students of Hons. and Post–Graduate classes of Indian Universities. This book has been written in such a way that in addition to the fundamentals, other important aspects have also been covered so far. Descriptions from *Cyclostomes* to *Mammals* in the vertebrate series, and, selected *Topics in Evolution* have been incorporated in this book, which are very useful for the students reading Zoology in Degree Colleges and Universities all over India. I hope that the book will be liked by the students and teachers of Zoology. Suggestions are always invited for the improvement of the book. We are pleased to record thanks to the publisher *Daya Publishing House*, Delhi for requesting to write this book; and publishing the same at the earliest.

— Authors

Contents

Chapter 1

The Chordata

The chordates form a large animal group which comprises *vertebrates* or animals with back bones belonging to Group – Craniata *(Vertebrata)* and three sub-phyla of animals i.e. *Hemichordata, Tunicata* or *Urochordata* and *Cephalochordata*. These three sub–phyla include animals which resemble vertebrates in certain respects though the vertebrae are lacking. These three sub-phyla are collectively known as *Protochordates* or *Lower chordates*. The Vertebrates are called higher chordates.

All chordates at some stage of the life-history possess *gills*, a dorsal *nerve cord* and a skeletal rod known as *notochord*. The *notochord* is also known as *Chorda dorsalis* and it is formed as longitudinal out growth from the archenteron. It is an elastic rod like structure situated below the nerve cord and above the alimentary canal. It is made up of vacuolated cells and covered with connective tissue sheath. It serves as skeleton providing rigid axis to the body of the animals concerned. In certain lower chordates it may persist in the adult condition also (e.g. in Amphioxus), in certain forms it is present in the larval form only (e.g. Herdmania). But in higher chordates it is seldom present in the adult stage. This structure contributes in the formation of vertebral column.

A dorsal tubular nerve cord is present in all the chordates. It forms the central nervous system. It modifies to form various structures later on. The whole nervous system is ectodermal in origin. The cavity of the original tubular structure is called the *neurocoel* at the time of development. Degeneration can take place in certain lower chordates.

Paired gill-slits are present in the wall of the Pharynx in all the chordates at some stage of the life history. The gill-slits are lined with vascular gill lamellae for respiration purpose. They persist throughout the whole life in lower chordates, fishes and some amphibians but in rest higher chordates they are present only during the developmental period.

In addition to the above basic characters, *bilateral symmetry, triploblastic condition, coelom* and *metamerism* can also be seen in the chordates, though these are present in some non-chordates also. Some other distinctive characters of the chordates are the presence of paired appendages, endoskeleton and exoskeleton, post anal tail, ventral heart, closed vascular system, Red blood anal cells, hepatic portal system, dorsal and ventral nerve roots, endocrine glands, mesenteries etc. Of course, these characters may not reach full development in all the chordates and may also be reduced in the later stage.

Origin of Chordates

So far as the origin of chordates is concerned, there are several theories which have been described below briefly. It is a fact that the chordates have originated from invertebrates and as such, there is scarcity of fossil records.

1. Coelenterate Theory

According to this theory, the chordates originated directly from the coelenterates. It has been argued that during the process of evolution, some specialized characters e.g. radial symmetry, nematocyst etc. were lost. Some advanced characters like bilateral symmetry, mesoderm segmentation, coelom etc. appeared during the course of evolution. But, these advanced characters are found in several invertebrates. So, this affinity has not been accepted by all zoologists.

2. Nemertine Theory

According to this theory, the marine worm like creature known as nemertine has been suggested as the ancestor of chordates on the basis of following characters :

(a) The medium dorsal nerve has been supposed to be the spinal cord of the vertebrates.

(b) The proboscis sheath present above the gut is comparable to the notochord of the chordates.

(c) The cerebral ganglion corresponds to the brain of the vertebrates.

(d) Lateral nerve indicates the lateral line of fishes.

(e) The cephalic pits on the sides of the head are comparable to the gill-slits.

But the nemertines are lowly organised forms. In some of the features e.g. ciliated epidermis, mesenchyme filled perivisceral space, flame cells and absence of respiratory organs, resemble more with the tubellarians rather than chordates. Also, proboscis sheath and cephalic pits are quite different from notochord and gill-slits. Thus, this nemertine theory had practically no support.

3. Annelid Theory

Because of certain resemblances between annelids and chordates, this theory has been proposed. This theory was originally supported by Dohrn (1875) and Minot (1897). The resemblances are —

(a) Bilateral symmetry and segmentation,

(b) Segmental excretory organs,

(c) Fully developed coelom,

(d) Closed circulatory system with longitudinal blood vessels.

But, the nerve cord in annelids occupies ventral position, also, it is a solid structure. Moreover in annelids, there is nothing like notochord or gill-slits. The pattern of development in both the groups are different. The annelid theory has been finally discarded.

4. Arachnid Theory

That the chordates have originated from Palaeozoic arachnid (Eurypterids) belonging to phylum – Arthropoda has also been advocated. This theory has been supported by Gaskell (1896) and Patten (1912). This theory is based on the resemblances between eurypterids and the extinct ostracoderms. Both were armoured with dorsal exokeleton. But, there is no arthoropod–like appendages in chordates. In arthropods the nerve cord is ventral and solid unlike that of chordates whereas it is dorsal and tubular structure. There is no reasonable ground to favour this view and as such, this theory was not supposed to be of any significance.

5. Echinoderm Theory

Now, zoologists consider that the echinoderms are the nearest relatives of chordates. Resemblances which have been shown between chordata and Echinoderms indicating relationship include :

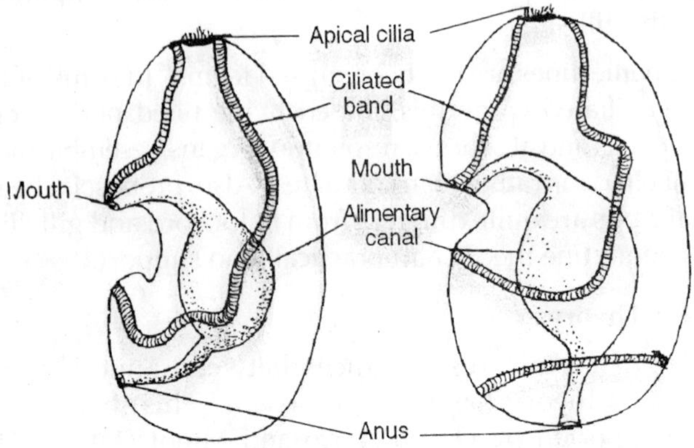

Fig.1 : *Tornaria* larva of hemichordate (right) and Echinoid larva (left) of echinoderm

(i) *Embryological evidence:* Both the echinoderms and chordates are the entercoelous animals. The free swimming bilaterally symmetrical *Bipinnaria* larva of Echinoderms resembles the *tornaria* larva of Balanoglossus (hemichordata). Muller (1950) thought that the tornaria larva was the echinoderm larva because of so much similarities. In both the larvae, microscopic oval and transparent body with apical tuft of sensory cilia, identical external ciliated bands, complete alimentary canal, five coelomic cavities and mode of development of central nervous system from ectodermal layer can be observed.

The embryos of lower chordates (e.g. Balanoglossus, Amphioxus) produce mesoderm in the from of paired coelomic pouches from the archenteron like echinoderm. The early history of cleavage, gastrulation and formation of the anus in the position of blastopore is almost the same in the above two phyla. Actually, the anus arises from the blastopore.

(ii) *Serological evidence*: Serological comparisons show a quantitative likeness between the echinoderms and chordates. Thus, a chemical similarity is there. The body-fluid proteins are very much similar. The phosphogens utilized in muscle contraction are similar in both the groups. The phosphogenes are the compounds of *creatine* and *arginine*. It may be pointed out that the phosphogenes in other invertebrates are the compounds of arginine only. Thus, chordates are supposed to be more related to echinoderms than any other group of invertebrates.

(iii) *Biochemical evidence:* By the action of *creatine phosphate*, the supply of ATP is maintained, till all the creatine phosphate is broken down. This happen in both the phyla (Echinodermata and Chordata). So, there is definitely some sort of ancestral link between the two phyla.

6. Neotenous Larva Theory

This theory was proposed by Gargstang (1894). It was shown that if the ciliated bands on the *auricularia larva* of holothurian (echinoderm) were accentuated and were facilitated to form ridges between which a groove was formed, and when the ridges fused, the groove was converted into a tube. This groove could be supposed to be the neurenteric canal and the tube was related to vertebrate nervous system. It was further presumed that some of the tailed larvae developed gonads precociously and developed into neotenous type of adult and did not metamorphose into sedentary adult. These neotenous adults invaded estuaries and river mouths for searching food. Among the palaeozoic animals, only protochordates and arthropods succeeded in exploiting the freshwater habitats. Thus, this theory advocates that the chordates probably evolved from the sexually mature *neotenic auricularia larva*.

Berril (1955) suggested the larval sequence in the way stated below :

Auricularia → Tornaria → Ascidian tadpole →

Permanently free-swimming chordate.

Here the ascidian tadpole has been considered significant in giving rise to chordates.

Conclusion

Echinoderms are supposed to be the nearest relatives of chordates. Resemblances between the two groups are such, that they indicate the evolutionary significance. Early Ostracoderms (e.g. *Drepanaspis*) resembled several carpoid echinoderms and as such, the evolutionary connection can not be ignored. It has also been suggested that chordates and echinoderms have been derived from more than one line leading back to forms not definable as either chordates or echinoderm. As there is lack of fossil records, the chordate origin is still a matter of conjecture.

Condensed Classification of Chordata

A brief classification of Phylum–chordata for fundamental concept, accepted by all has been presented below in condensed form :

Phylum–chordata has been divided into two groups i.e. *Acrania* and *Craniata*. Acrania is also known as Protochordata and Craniata is called vertebrata. In Acrania, there is no cranium or brain box, Jaws and brain but in Craniata the brain box or cranium, Jaws and brain are necessarily present.

Group – Acrania (Protochordata)

Sub–Phylum — *Hemichordata*
Class — Enteropneusta e.g. *Balanoglossus*
Class — Pterobranchia e.g. *Cephalodiscus, Rhabdopleura*
Sub–Phylum — *Urochordata* (Tunicata)
Class — Larvacea e.g. *Oikopleura*
Class — Thaliacea e.g. *Salpa*
Class — Ascidiacea e.g. *Herdmania*
Sub–Phylum — *Cephalochordata* e.g. *Branchiostoma* (Amphioxus)

Group – Craniata (Vertebrata)

Division — Agnatha (No true jaws and paired appendages)
Class — *Cyclostomata*
Sub–Class — Cephalaspidomorphi
Order – Anaspida e.g. *Birkenia*
Order – Osteotraci e.g. Ostracoderm (extinct)
Order – Petromyzontia e.g. Petromyzon
Sub–Class — Pteraspidomorphi

Order – Heterostraci e.g. Ostracoderm
Order – Myxinoidea e.g. Hag fishes

Division — *Gnathostomata* (With true jaws and paired appendages)

Super Class — Pisces
Class — *Placodermi*
Order – Acanthodii e.g. *Climatius*
Order – Arthorodira e.g. *Coccosteus*
Order – Antiarchi e.g. *Pterichthyodes*
Class — *Chondrichthyes (Elasmobranchi)*
Order – Cladoselachii e.g. *Cladoselache* (shark)
Order – Selachii e.g. *Hexanchus* (shark)
Order – Batoidei e.g. *Raja* (Ray)
Class — *Holocephali*
Order – Chimaerae e.g. *Chimaeras*
Class — *Osteichthyes (Teleosti)*
Sub–Class — Actinopterygi (Ray-finned fishes)
Super–Order — Chondrostei e.g. Polypterus, sturgeons
Super–Order — Holostei e.g. Gars, bow fins
Super–Order — Teleosti (Modern bony fishes) .
Class — *Choanichthyes*
Order – Dipnoi e.g. Lung fishes
(Now, Dipnoi has been given the status of independent class)
Order – Crossopterygi e.g. Lobe finned fishes
Sub–Order – Rhipidistia e.g. *Eusthenopteron*
Sub–Order – Coelacanthini e.g. *Latimeria chalumnae*
Super Class — TETRAPODA
Class — *Amphibia*
Sub–Class — Apsispondyli
Super Order — Labyrinthodontia
Order – Ichthyostegalia e.g. Eryops
Order – Rhacitomi
Order – Stereospondyli e.g. Capitosaurus
Order – Embolomeri e.g. Palaeogyrinus
Order – Seymouriamorpha
Super–Order — Silentia
Order – Anura e.g. Frogs and Toads
Super–Order — Caudata
Order – Urodela e.g. *Salamander, Proteus, Necturus*

Sub–Class — Lepospondyli
Order – Microsauria e.g. Microbronchus
Order – Apoda (Gymnophiona) e.g. *Caecilians*
Sub–Class — Phyllospondyli (Salamander like small amphibians, though considered as imaginary group by Romer, 1945).
Class — *Reptilia*
Sub–Class — Anapsida (Skull complete without temporal vacuity or fossa).
Order – Cotylosauria e.g. *Seymouria*
Order – Chelonia e.g. Turtles and Tortoises
Sub–Class — Parapsida (Skull with one pair of temporal vacuity, one on either side)
Order – Protosauria e.g. Araeoscelis
Order – Pleistosauria e.g. Placodus
Order – Ichthyosauria e.g. *Ichthyosaurus*
Sub–Class — Diapsida (Two temporal vacuities on each side of the skull)
Super–Order — Lepidosauria
Order – Eusuchia e.g. *Youngina*
Order – Rhynchocephalia e.g. *Sphenodon*
Order – Squamata
Sub–Order – Lacertilia e.g. Lizards
Sub–Order – Ophidia e.g. Snakes
Super–Order — Archosauria
Order – Thecodontia e.g. *Aetosaurus*
Order – Crocodilia e.g. Crocodiles, Alligators
Order – Saurischia e.g. Dinosaurs (*Brontosaurus*)
Order – Ornithischia e.g. *Stegosaurus*
Order – Pterosauria e.g. Pterodactyls
Sub–Class — Synapsida (Only one pair of temporal vacuities, one on each side of skull)
Order – Pelycosauria e.g. *Dimetrodon*
Order – Therapsida e.g. Mammal like reptiles (*Bicynodon*)
Class — *Aves* (Includes Birds)
Sub–Class — Archaeornithes e.g. *Archaeopteryx*, *Archaeornis*
Sub–Class — Neornithes
Super–Order — Odontognathae e.g. *Hesperornis, Ichthyornis*
Super–Order — Palaeognathae (Flightless, running birds)
Order – Struthioniformes e.g. *Struthiocamelus* (Ostrich)

Order – Casuariformes e.g. *Dromaius* (Emu)
Order – Tinamiformes e.g. Tinamus
Order – Rheiformes e.g. *Rhea*
Order – Apterygiformes e.g. *Apteryx* (Kiwi)
Order – Aepyornithiformes e.g. *Aepyornis*
Order – Dinornithiformes e.g. *Dinornis*
Order – Spehnisciformes e.g. *Penguin* (Aptenodytes)

Super–Order — Neognathae – Flying birds (Carinatae)

Order – Gaviiformes e.g. Loons (*Gavia* sp.)
Order – Colymbiformes e.g. Grebes
Order – Procellariformes e.g. Petrels
Order – Pelicaniformes e.g. Pelicans, Cormorants
Order – Ciconiiformes e.g. storks and herons (Ardea)
Order – Anseriformes e.g. Ducks, Geese, Swans
Order – Falconiformes e.g. Hawks and Falcons
Order – Galliformes e.g. Game birds (Turkeys, *Gallus* etc.)
Order – Gruiformes e.g. Cranes (*Grus*)
Order – Charadriiformes e.g. waders and gulls
Order – Columbiformes e.g. Pigeons (*Columba*)
Order – Cuculiformes e.g. Cuckoos (*Cuculus*)
Order – Podicipitiformes e.g. Podiceps
Order – Psittaciformes e.g. Parrot
Order – Strigiformes e.g. Owls
Order – Caprimulgiformes e.g. *Caprimulgus*
Order – Micropodiiformes e.g. Swifts and Humming birds
Order – Coraciiformes e.g. Bea eaters and King fishers
Order – Piciformes e.g. Wood peckers
Order – Passeriformes e.g. *Passer domesticus*
Order – Trogoniformes e.g. Trogons
Order – Coliiformes e.g. *colius*

Class — *Mammalia*

Sub–Class — Prototheria (Egg laying mammals)

Order – Monotremata e.g. Echidna *(Tachyglossus)*
Order – Triconodonta e.g. *Triconodon*
Order – Symmetrodonta e.g. Peralestes

Sub–Class — Allotheria

Order – Multituberculata e.g. *Ctenacodon*

Sub–Class — Theria
Infra–Class — Pantotheria
Order – Dryolestoidea e.g. *Amphitherium*
Order – Docodonta e.g. *Docodon*
Infra–Class — Metatheria
Order – Marsupialia e.g. Kangaroo
Infra–Class — Eutheria (Placentalia)
Order – Insectivora e.g. *Talpa*
Order – Dermoptera e.g. *Galeopithecus*
Order – Chiroptera
Sub–Order – Microchiroptera e.g. *Scotophilus*
Sub–Order – Megachiroptera e.g. *Pteropus*
Order – Primates
Sub–Order – Lemuroidea e.g. *Lemur, Loris*
Sub–Order – Tarsioidea e.g. *Tarsius spectrum*
Sub–Order – Anthropoidea
Family – Cebidae (New World Monkeys)
Family – Cercopithecidae (Old World Monkeys)
Family – Callithricidae e.g. *Callithrix*
Family – Pongidae e.g. Apes
Family – Hominidae e.g. Man
Order – Edentata e.g. *Armadillo*
Order – Pholidota e.g. Pangolins *(Manis)*
Order – Lagomorpha e.g. Rabbit, Hares
Order – Rodentia e.g. Mice, Squirrel (*Funambulus*)
Order – Cetacea e.g. Whales
Order – Carnivora e.g. Cats, dogs, lions, tigers, etc.
Order – Condylarthra e.g. *Phenacodus*
Order – Proboscidea e.g. Elephants (*Elephas*)
Order – Tubulidentata e.g. *Orycteropus*
Order – Hyracoidea e.g. *Hyrax*
Order – Sirenia e.g. Manatee (Sea cows)
Order – Perissodactyla e.g. Horses *(Equus)*, Rhinoceros, Zebra
Order – Artidactyla e.g. Pig (Sus), deer, goat, etc.

Chapter 2

Class — Cyclostomata

The cyclostomata (Gr. Cycl, Circular; Stom, Mouth) includes the agnathans, the Jawless vertebrates. They are the only vertebrates without jaws. The fossil records indicate that the cyclostomes evolved during the devonian period. They are modified and degenerate offshoot of the primitive vertebrate stock (obstracoderm stock). Once the *Ostracoderms* were abundant thoughout the world but they became extinct, and only two types of cyclostomes survive as relics of these agnathan vertebrates. The surviving forms are the lampreys (e.g. Petromyzon) and hagfishes (e.g. Myxine). They have rounded bodies and funnel shaped sucking mouths and they are parasites on other fishes. They have soft cartilaginous skeleton and no paired fins. They have only one nostril and no scale. They have laterally compressed protocercal tail. There is no scale over the body. Gills are present in the muscular pouches arranged in the linear fashion. Opening through pores in each side there are 6-16 pairs of gills in different species. They have both primitive and specialised characters. Median fins with cartilaginous fin rays are present. One or two semi-circular canals are present in each auditory organ. A muscular tongue is present which bears epidermal teeth, an adaptation for rasping the flesh of the fishes.

The class–Cyclostomata includes two orders which include the living forms i.e. (1) *Petromyzontia* (example – *Petromyzon*), (2) *Myxinoidea* (example – *Myxine*). Animals belonging to orders *Anaspida* and *Osteostraci* (both belonging to sub class – *Cephalaspidomorphi*) have already become extinct. The animals belonging to order–*Heterostraci* (Sub class – *Pteraspidomorphi*) have also become extinct. Petromyzon and Myxine belong to sub class – *Cephalaspidomorphi* and *Pteraspidomorphi* respectively.

Comparison Between Lampreys and Hagfishes

***Lampreys** (e.g. Petromyzon)*	***Hagfishes** (e.g. Myxine)*
1. They are both marine and freshwater forms. They normally migrate to rivers for spawning.	1. They are exclusively marine forms and they remain buried in sand.
2. They have cylindrical and stout body.	2. The body is 'eel like' slender and feeble.
3. The larval form (e.g. ammocoete larva) leads an independent life and the adults are semiparasite (ectoparasite).	3. They are parasites on larger fishes and they may enter inside the bodies of the prey.
4. The mouth is subterminal forming a funnel like structure (buccal funnel).	4. Mouth is terminal in position surrounded by soft lips and there is no funnel.
5. There is no tentacle around the mouth.	5. There are four pairs of tentacles around the mouth.
6. The dorsal fin is divided by a notch. The caudal fin is well marked.	6. The dorsal fin is poorly developed however the caudal fin is 'slightly' developed.
7. There is a single nostril on the dorsal side, occupying a median position.	7. There is a single nostril lying very close to the mouth and opens terminally.
8. The tongue is feebly developed having larger teeth.	8. The tongue is strongly developed having smaller teeth.
9. The skull is imperfectly roofed.	9. The skull is without roof.
10. Horny teeth are present in the buccal cavity and tongue. The teeth on the tongue are larger in size.	10. Small sized teeth are present on the tongue in two rows.
11. Seven pairs of gill-slits are present in the adult.	11. Only one pair of gill-slits are present in Myxine, but in some forms (e.g. *Eptatratus* there may be 13-14 pairs).
12. Paired salivary glands are present. Its secretion is anticoagulant which prevents coagulation of blood of the host.	12. Paired salivary glands are normally absent.
13. Gill-pouches (seven pairs) open into the respiratory tube, not into the gut.	13. Gill pouches (6 pairs) open directly into the gut.
14. The afferent and efferent branchial arteries supply each the posterior hemibranch of one gill pouch and the anterior hemibranch of the next one.	14. Each aortic arch supplies the hemibranch of a single gill pouch.
15. Paired eyes are well developed and functional.	15. Eyes are vestigial and rudimentary, eye muscles and lenses are absent. It is because of bottom dwelling habit.

Lampreys (e.g. Petromyzon)	Hagfishes (e.g. Myxine)
16. The brain is well developed.	16. The brain is degenerated.
17. 10 pairs of cranial nerves are present.	17. Always less than 10 pairs of cranial nerves. Only oculomotor (III), trochlear (IV), and abducens (VI) are prominent.
18. Special value is present in the intestine.	18. Longitudinal folds are present in the intestine, there is no spiral valve.
19. Urino–genital sinus is present.	19. The urinogenital sinus is absent.
20. The segmentation is holoblastic.	20. The segmentation is meroblastic.
21. The development is indirect because the ammocoete larva undergoes metamorphosis.	21. Development is direct without larval form and its metamorphosis.
22. Nasohypophysial duct ends blindly.	22. Nasohypophysial duct opens posteriorly in the mouth cavity.
23. Eggs are small and without horny shell.	23. Eggs are large with horny, shell and hook like processes.
24. The pronephros does not persist in the adult forms. The mesonephros is the functional adult kidney.	24. The pronephros is retained in the adult forms. The mesonephric kidney is functional.
25. The sexes are separate but differentiation takes place in the adult stage.	25. The sexes are united. The anterior part of the gonad acts as ovary and the posterior parts as testis (hence called Ovotestis).
26. They are anadromous and during the breeding season they migrate to freshwater (rivers).	26. They are purely marine and spawning takes place in the sea.

Salient Features and Affinities of Cyclostomata

(Example – Petromyzon)

Structure: *Petromyzon marinus* has almost world–wide distribution. They live in the sea and migrate to the rivers for spawning in autumn, while the sexual maturity is attained in winter. The breeding occurs in the spring season. The important morphological and anatomical features have been described below :

1. The body of the adult Petromyzon is elongated, cylindrical and eel-like. The *Petromyzon marinus* may be up to 1 metre long. The *Lampetra fluviatilis* is about 90 cm. long and *L. planeri* is only about 45 cm. long.
2. The median *dorsal fin* is divided into two unequal parts by a notch. The caudal fin is continuous with posterior

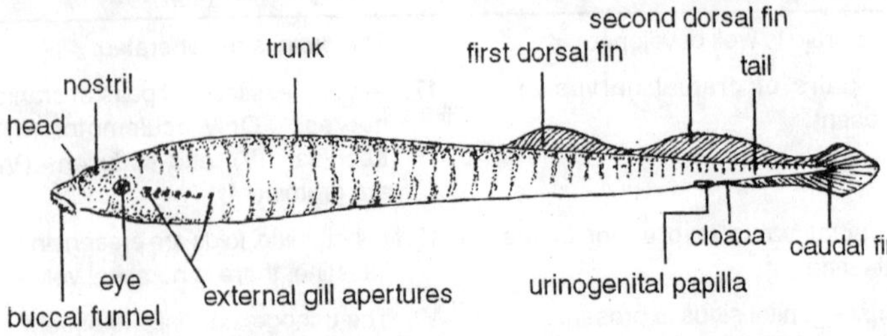

External features of Petromyzon marinus

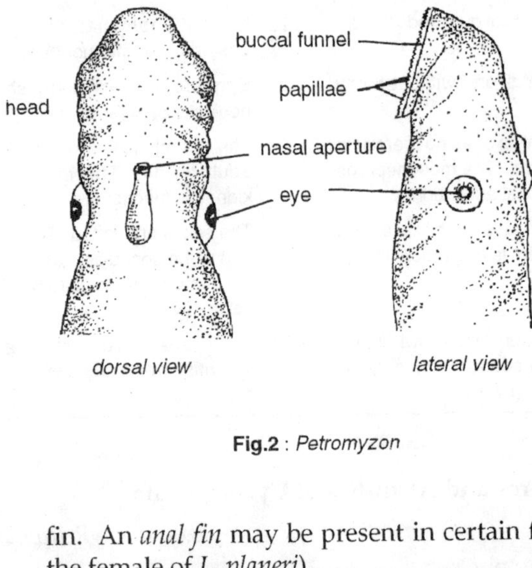

Fig.2 : *Petromyzon*

fin. An *anal fin* may be present in certain forms (e.g. in the female of *L. planeri*).

3. The whole body can be conveniently divided into three regions viz. Head, Trunk and Tail. The tail portion is more or less laterally compressed.
4. There is no scale, and paired fins are absent.
5. The *anus* is present at the junction of the trunk and tail in the ventral side.
6. There is a papilla behind the anus and at its tip the *urino-genital aperture* is present. The anus and urinogenital aperture open separately into a cloacal pit. There is no true cloaca.

7. In the male individual the penis is present which is an eversible structure.
8. In the head region the *buccal funnel* is situated anteriorly which is directed downwardly.
9. At the bottom of the buccal funnel, the mouth is present which is a circular opening.
10. In the peripheral area of the buccal funnel, a membrane is present. From this membrane arise the *Oral fimbriae* and *sensory cirri*. Sensory cirri are longer structures.

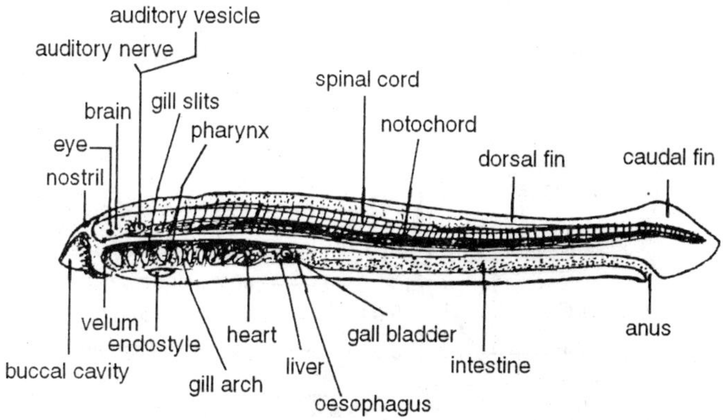

Fig. – *Ammocoete larva of Petromyzon*

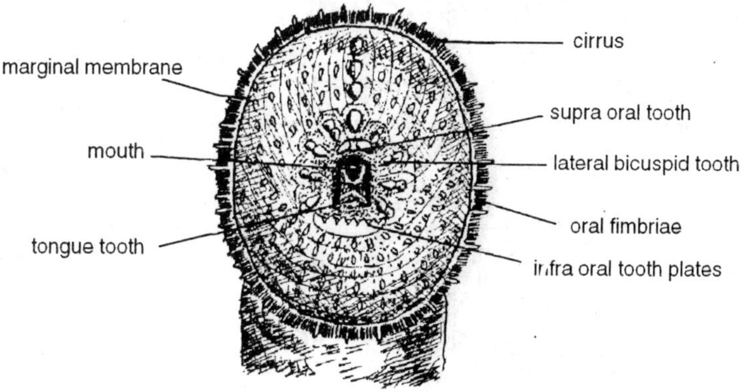

Fig. – *Petromyzon*-buccal funnel

Fig.3 : *Petromyzon — Ammocoete larva,* buccal funnel

11. In the buccal funnel radiating rows of *horny teeth* are present which rest on cartilaginous pads.
12. From the bottom of the buccal funnel, the tongue projects and it is provided with horny teeth.
13. There is a single median *nostril* on the dorsal side of the head. Behind the nostril, the transparent area of the skin is the position of the pineal organ.
14. The paired eyes are without eye-lids and covered with transparent skin.
15. In the postero–lateral sides of the head, seven pairs of *gill–slits* are present.
16. Lateral *sensory pits* are present in the lateral sides of the head and tail.

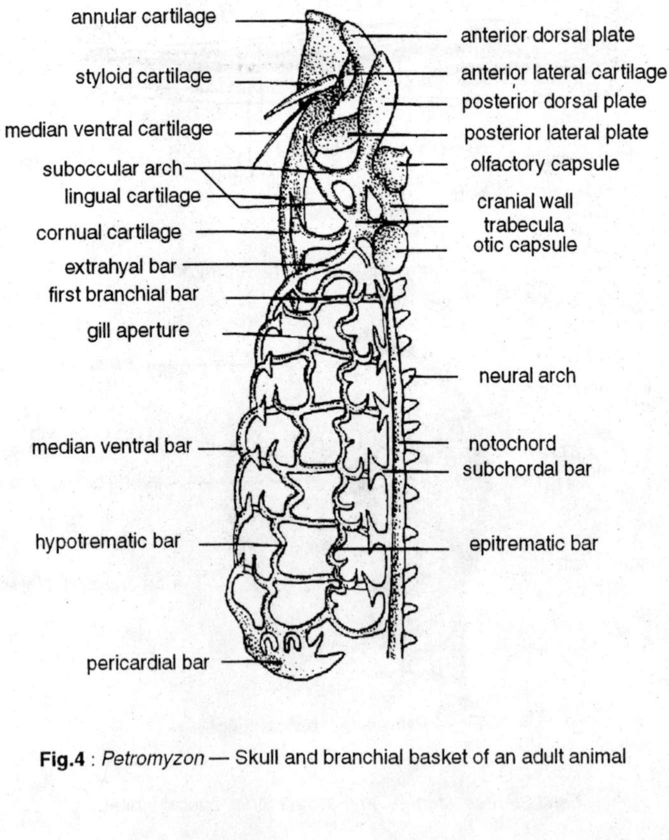

Fig.4 : *Petromyzon* — Skull and branchial basket of an adult animal

17. The skin consists of *epidermis, dermis,* and *subcutaneous* layer. Chromatophores are present in both the epidermis and dermis.

18. *Myomeres* are prominent in the trunk and tail regions and they are separated by *myocommas.*

19. The notochord is persistent. The skull is a primitive structure. The basal plate of the floor is formed by the union of *parachordals* and *trabeculae* which surround the anterior end of the notochord. Several pieces of cartilages are attached with the basal plate to form the cartilaginous box to enclose the brain and the special sense organs. There is no roof except a narrow occipital arch (*Synotic tectum*) between the otic capsules. Posterior ends of the trabeculae do not unite and as such, a large aperture is formed, known as *basi–cranial fontanelle.* The *naso–hypophysial sac* passes through this aperture. The unpaired olfactory capsule unites with the cranium by fibrous tissue. The sub–ocular arch supports the eye in both the sides of the basal plate. A slender styloid process hangs from the subocular plate which is connected with *cornual cartilage.*

20. The visceral skeleton is present in the form of *branchial basket* which supports the outer part of the branchial passage. Contraction and expansion of branchial basket helps the process of respiration. The basket is made up of cartilages and rods. There are nine vertical bars connected with each other by four longitudinal bars. The first vertical bar is present posterior to the *styloid cartilage* and the second one lies in front of the first gill–cleft. The last vertical bar is connected with a cup–like pericardial cartilage to accommodate the heart. The other important longitudinal bars are *Median vertical bar, Epitrematic bars, Hypotrematic bars* and *Sub–chordal bar.* The buccal funnel is supported by *annular cartilage* and the tongue has the support of *lingual cartilage.*

21. The mouth is situated in the buccal funnel and leads into the buccal cavity. The buccal cavity communicates with two tubes – the dorsal tube is the *oesophagus* and the ventral tube is the respiratory tube (Pharynx). The entrance of the respiratory tube is guarded by *velum* provided with velar tentacles. When the animal feeds, it becomes the

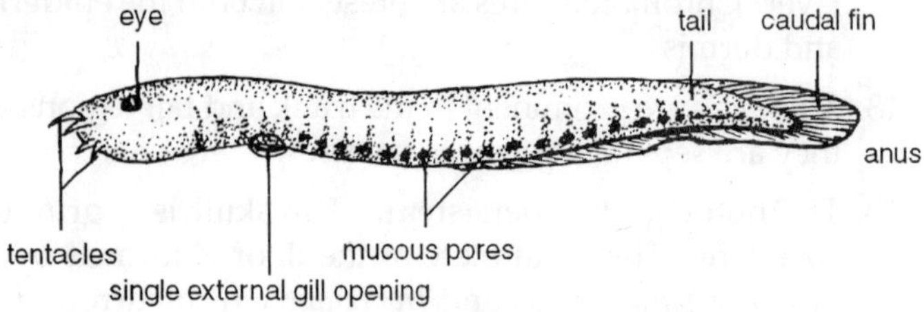

Fig. – *Myxine glutinosa*

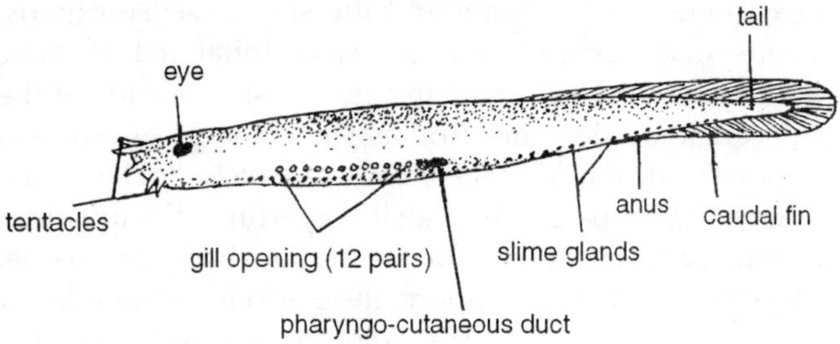

Fig. – *Bdellostoma*

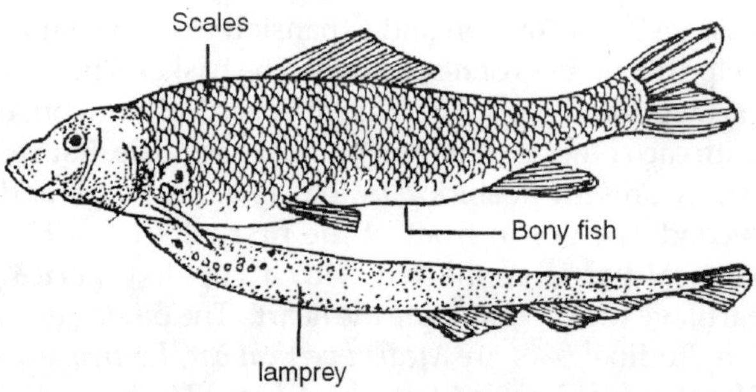

Fig. – *Lamprey attached to a bony fish*

Fig.5 : Some Cyclostomes

functional duty of the velar tentacles to keep the digestive tube completely separated from the respiratory tube. The oesophagus leads into a straight intestine. There is no true intestine. The posterior part of the intestine forms the rectum and the latter is connected to a cloaca. The typhlosole forms the spiral value in the intestine by following a spiral course. The liver and gall bladder are well developed and there is no pancreas. Gut epithelial cells contain zymogen cells and follicles of Langerhans. These cells secrete enzymes for digestion and they are also responsible for carbohydrate metabolism.

22. There are seven pairs of gill pouches (gills) with a number of gill lamellae. They open outside by external gill slits and inside by internal gill slits. Gill pouches are separated from each other by inter-branchial septa. By the contraction and expansion of branchial baskets, water passes through the gill lamellae and O_2 is thus absorbed.

23. The heart is S-shaped. It consists of three chambers – one *Sinus Venosus*, one *auricle* and one *ventricle*. The sinus venosus is attached with the auricle and the latter opens into the ventricle. There is no *conus arteriosus*, the dilated part of the ventral aorta is known as *bulbous arteriosus*. The pericardium, supported by cartilaginous plate encloses the heart. Lymphalic channels are absent in lamprey. Afferent and efferent branchial arteries supply each the posterior hemibranch of one gill-pouch and the anterior hemibranch of the next one. There is no renal portal system however, the hepatic portal system is present. The left precaval is absent in the adult. The jugular and cardinals of both the sides open into the right precaval. The hemoglobin is present in the blood.

24. The brain is well developed and comprises fore-brain, mid brain and hind brain. The fore brain includes paired olfactory lobes. The cerebral hemispheres are smaller and attached to the diencephalon. On the dorsal side of diencephalon, there are two unequal habenular ganglia. The pineal or median eye lies dorsal to the infundibulum. The hind brain is differentiated into cerebellum and medulla oblongata. The choroid process is present between the optic lobes and the medulla oblongata. The spinal cord is dorso-ventrally flat. The sympathetic nervous system is poorly developed.

25. The functional adult kidney is of mesonephric type which develops from the nephorgenous tissues. The mesonephric duct opens into the urinogenital sinus behind the rectum. It opens by a urinogenital papilla into the cloaca in which the alimentary canal also ends.

26. The paired eyes are large in size and they are functional. In addition to these, there are two median eyes also on the top of the brain, known as the pineal and parietal eyes. The parietal eye is present beneath the pineal eye and it is rudimentary. There are two semicircular canals in the ears. The lateral line sense organs are simple occurring in

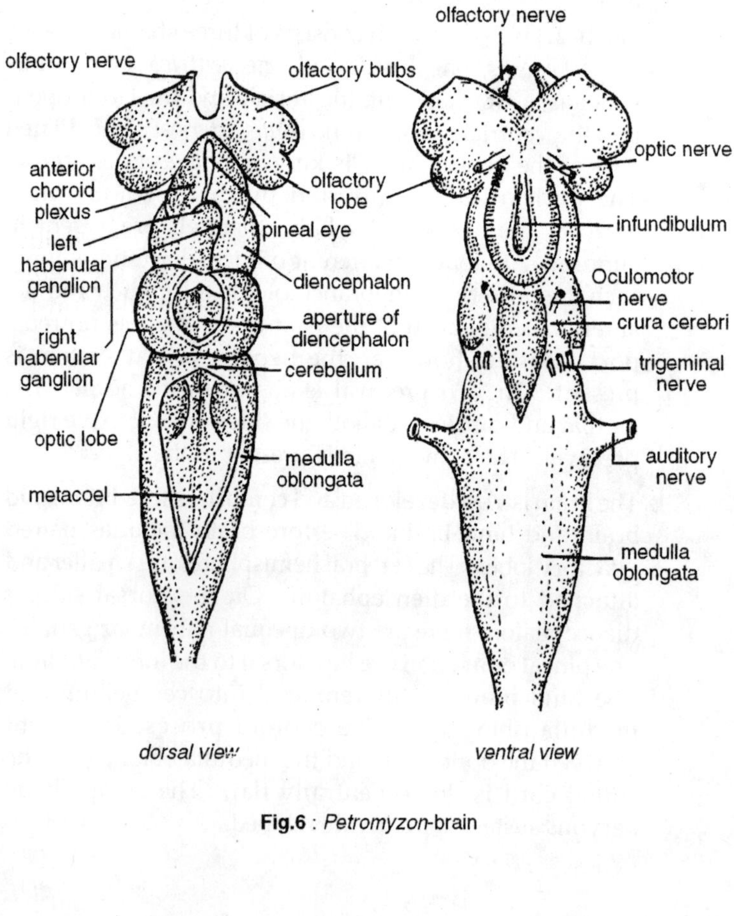

Fig.6 : *Petromyzon*-brain

the head region and extending on single line along each side of the body.

27. The sexes are separate in the adults. In the young petromyzon the male and female sex cells are present in the same gonad. The genital ducts are absent. The ova and sperms are liberated into the coelom by the rupture of gonads. The gametes come out through a pair of genital pores. Fertilization is external.

28. After spawning, the fertilized eggs are embedded in the mud/sand of the nest made in the river beds. The eggs are telolecithal but the cleavage is holoblastic. After two to three years the young one hatches from the egg and it is known as *ammocoete larva*. At first it is only 1 cm. long but later on it reaches a length of up to 170 cm. The larva feeds on algae and bacteria. During the process of feeding, the muscular movement of velum and pharynx takes place. The larval period continues for 3-7 years and then the larva metamorphoses into the adult. The endostyle becomes modified to form the thyroid gland.

Affinities

Cyclostomes are the living forms, derived from the primitive vertebrate stock (Ostractoderm). They have primitive, specialized and degenerate characters which have been discussed below :

1. Primitive Characters

Characters resembling Branchiostoma

(a) Persistent notochord

(b) Numerous gill-slits

(c) Segmental muscles (myotomes) with some modifications

(d) Straight and ciliated alimentary canal without much regional specialization

(e) Absence of gonoducts

(f) Fertilization is external

(g) Head is not distinct

(h) Paired fins absent

(i) Development is indirect as there is a larval form which metamorphoses into adult form. The *ammocoete larva* resembles Amphioxus in general appearance and also in having endostyle, continuous median fin, ciliated gut and filter feeding etc.

Differences from fishes

(a) Absence of scales, paired fins, hinged jaws and true teeth

(b) Continuous median fin-fold

(c) Absence of gall bladder, gonoducts, spleen, girdles and ribs

(d) Diphycercal caudal fin

(e) Incomplete cranium

(f) Vertebrae either absent or poorly developed

(g) Pancreas is rudimentary

(h) There is a single median nostril

(i) Heart is S-shaped and there is no conus arteriosus

(j) Only one or two semi circular canals in the ear

(k) Brain is smaller in size

(l) Lateral line system is poorly developed

(m) Sympathetic nervous system is poorly developed

Affinities with Ostracoderms

Resemblances

(a) Absence of jaws

(b) Brain structure

(c) Gills in muscular pouch

(d) Arrangement of cranial nerves

(e) Nasal opening single

(f) Two semi circular canals in the auditory organ

(g) Lateral line system

Differences

(a) In ostracoderms there were dermal scales on the trunk and tail

(b) Bony skeleton was present in the anterior part of the body

(c) All ostracoderms were no parasites

(d) Some ostracoderms had a pair of pectoral fins.

2. Vertebrate Characters

(a) Head region has paired eyes and ears

(b) Gills are present for respiration

(c) Cranium present for the protection of brain

(d) 10 pairs of cranial nerves like fishes, amphibians and birds.

(e) Bilobed large liver

(f) Red blood corpuscles and white blood corpuscles present

(g) Well developed circulatory system with a heart present

(h) Kidney is that of mesonephric type

(i) Lateral line sense organs present

(j) Myotomes like fishes present

(k) Vertebrae incomplete however, beginning of segmental vertebrae is there.

(l) Dorsal roof ganglia on spinal nerves.

3. Specialized Characters

(a) Head and cranium present

(b) Brain is quite advanced

(c) Sucktorial mouth for parasitic life

(d) Pro – and mesonephric kidney

(e) Several layered epidermis

(f) Secondary notochord

(g) Sac like gill chambers far behind the head

(h) Respiratory pharynx separated from oesophagus

(i) Presence of branchial basket

(j) Presence of liver

(k) A single dorsal median nostril

(l) Large eggs with sufficient yolk, meroblastic cleavage

(m) Presence of tongue with teeth.

4. Degenerate Characters

(a) Elongated eel–like body

(b) Lack of gallbladder and bile duct

(c) There is no exoskeleton

(d) Absence of paired fins and girdles

(e) Eyelids are absent

(f) Lack of ossification of endoskeleton.

Conclusion

Thus cyclostomes have developed several characters which have been mentioned above. They differ from fishes in several important features like – absence of biting jaws, absence of paired fins, presence of branchial basket, presence of rasping tongue etc. Hence, placing of cyclostomes in *Agnatha* separating them from fishes, seems to be justified. According to Stensiö (1927) the cyclostomes have descended from the archaic stock, the *ostracoderms*.

There are evidences that *Cephalaspids* and *Pterapsids* are mainly concerned with the derivation of cyclostomes. The important characters of *cephalaspis* (Order – Cephalaspida) were —

(a) Between the eyes and behind the nostril, there was a pineal opening.

(b) On each side, there were 10 gill openings corresponding to 10 gill pouches that occupied each side of the head shield.

(c) Soft rays were present in the heterocercal tail.

(d) Paired fins joined the body on either side behind the horns of the head shield.

Similarly, the important features of Pteraspis (Order – Pteraspida) were —

(a) They were small and heavily armoured.

(b) Mouth was present in the form of a transverse slit placed near the front on the lower surface on the head shield. There was a series of slender plates stretched across the slit which might have acted like jaws in absence of true jaws.

(c) On each side, there was a single exit for the gills on the dorsal middle–line on the head shield, a long spine projected up and back.

(d) The tail was *reversed heterocercal.*

Stensiö (1927) has advocated a diphylatic origin of cyclostones. The lampreys have descended from the *cephalaspids* and the hag fishes have probably been derived from *pteraspids.* According to many authors the exact origin of cyclostomes is still doubtful.

Chapter 3
Pisces (Fishes)

Fishes are cold–blooded aquatic vertebrates. They normally breathe by means of gills, but sometimes accessory respiratory organs may be present. There are more than forty thousand species of fishes living in fresh water and marine water. They have stream-lined body but sometimes snake like forms are also there. Some are dorso–ventrally flattened which is an adaptational feature. They have paired (Pectorals and pelvics) and unpaired (Dorsal, anal and caudal) fins which are provided with soft and spiny fin rays. The integument is either naked or beset with scales, dermal denticles or bony plates. In the most primitive jawed fishes, there was *autostylic* jaw suspension, in the primitive sharks it was *amphistylic* and in the later sharks and bony fishes the jaw suspension is of *hyostylic* type. Paired nostrils are present which normally do not open into the pharynx except in lung fishes and lobe fin fishes. There is a lateral line system and receptors are well developed. The heart consists of only impure blood and hence it is known as the venous heart. It has one auricle and one ventricle. The Venous blood is pumped to the gills for aeration. The gills are covered by bony operculum in teleosts. The sexes are separate. The Kidney is of mesonephros type. There are 10 cranial nerves. The endoskeleton is cartilaginous (e.g. chondrichthyes) or bony (e.g. osteichthyes). Devonian period of the palaeozoic era was considered as the age of fishes of all types.

Dipnoi

The Dipnoi (double breathers) is an important group of bony fishes which evolved during the Devonian period. They are also known as "Lung–fishes" and actually represent a transitional group between fishes and amphibians. Several fossil forms like— *Dipterus* (Dipteridae), *Phaneropleuron, Dipnorhynchus, Uronemus*

(*Uronemidae*), and *Creatodus* (Ceratodontidae) are on record. But, the members of only three genera are surviving as present and they are *Neoceratodus, Protopterus* (Protopteridae) and *Lepidosiren* (Lepidosirenidae). There are four important species of *Protopterus*. They are *P. annectans, P. aethiopicus, P. dollei* and *P. amphibius*. Of the *Neoceratodus, N. forsteri* is a common species.

Class–Dipnoi comprises only two orders : *Monopneumona* (with single lung) e.g. *Neoceratodus*, and *Dipneumona* (with double lungs) e.g. *Protopterus* and *Lepidosiren paradoxa*.

The earliest fossil fish represented by the genus *Dipterus* was a fish of middle devonian age which possessed many of the generalized choanate characters like—long fusiform body, strong heterocercal tail, archipterygial type paired fins, two dorsal fins and large heavy cosmoid scale. Some spacialized features of this fish were : poorly ossified brain case composed of numerous bony plates, particularly ossified jaws, and dentition highly specialized with suppressed marginal teeth in both the jaws.

The central line of dipnoan evolution led to ceratodus which were distributed during the Triassic and subsequent period of the Mesozoic era.

Living Forms of Dipnoi

There are three living forms of the lung fishes included in the class Dipnoi. They have been described below :

(1) *Neoceratodus* : These are found in the Burnett and Mary rivers of Queensland (Australia). It has stout and cylindrical body and may attain a length of 6 feet. They are most primitive among the living forms and resemble the lung fishes of Triassic age. It has highly vascular lung. It can not live without water. They frequently come to the surface and breathe air. This fish is able to walk along the bottom of the rivers or ponds by using paired fins. It is a direct descendant of ceratodus. It has pointed head with large rounded scales all over the body. There is great reduction in the internal skeleton including the skull bones. Elongated and leaf shaped paired fins are present, the fleshy portion of which is covered with scales. Simplified tail is known as a *gephyrocercal* fin formed by fusion of original dorsal, caudal and anal fins.

(2) *Protopterus* : It is found in rivers and marshes of West

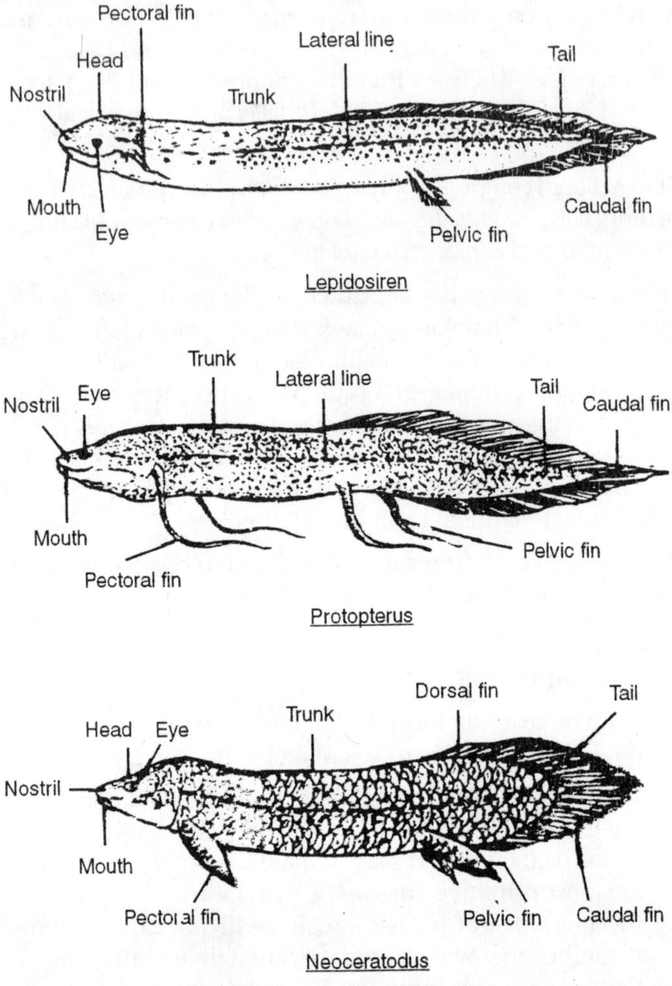

Fig.7 : Lung fishes (living)

Africa. This fish is less bulky than Neoceratodus. It can live for months without water. With the advent of dry season, it burrows into the mud leaving one or more openings on contact with the outer air. It has one pair of lungs. The paired fins are reduced to long slender whiplike appendages. They are characterized by *gephyrocercal* tail.

(3) *Lepidosiren* : They are found in the swamps of the Amazon basin (South America). They are eel-shaped creatures with small scales. It has two lungs. In dry season it also burrows in mud like *Protopterus*. There is great reduction of paired fins in this fish and as such they look smaller. Like *Protopterus* it has gephyrocercal tail showing evolutionary trends similiar to that of *Neoceratodus*.

General Characters of Dipnoi

(1) Paired fins are more or less acutely lobate and there are overlapping cycloid scales which are embedded in the dermis. In *Dipterus* the scales were covered with a layer of *cosmine*. In the living dipnoi paired fins are supported by jointed dermal fin rays which are bony in nature. The fins are of the archipterygial type. The tail is diphycercal with continuous dorsal and ventral median fins. *Dipterus* had a heterocercal tail, there were two separate dorsal and one anal fin.

(2) Two internal nostrils are present on the ventral surface of the skull and open into the roof of the mouth, facilitating aerial respiration as in frog.

(3) The skull is *autostylic* i.e. hyoid arch does not take part in the formation of suspensorium. Large dental plates are

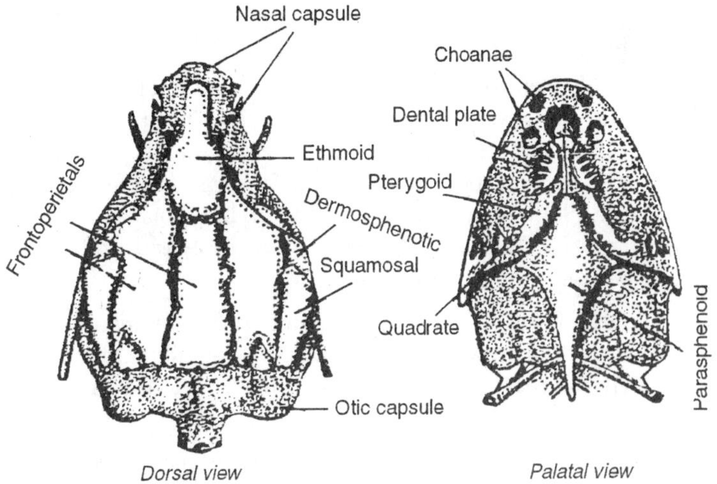

Fig.8 : Skull of *Neoceratodus*

firmly fused to the bones of the jaw. They are composed of thick dentine.

(4) There is only one external branchial aperture and there is absence of spiracles. The air bladder is highly developed to form a breathing lung.having cross–septa and alveoli. Lungs are well vascularized.

(5) There is a persistant and unconstricted notochord covered with a fibrous sheath. The skeleton is largely cartillaginous.

(6) In the skull, premaxillae, maxillae and the nasal are absent. The hyomandibulars are also absent. Dentaries are absent in the lower jaw and teeth are borne by the coronoids. There is great reduction in the ossification of the skull.

(7) Mouth is situated on the ventral surface of the head. The pharynx leads into the oesophagus. The oesophagus is connected with the stomach and the latter leads into the intestine. The rectum opens into the cloaca which opens by an aperture at the root of the tail. The liver is biloLed. The gall bladder lies between the two lobes of the liver. The pancreas is embedded in the intentina! wall. There is a well developed spiral valve in the intestine. Hepatic caecae are absent.

(8) In the brain single optic lobe is present. The cerebral hemispheres are quite large. The Pineal body is also present.

(9) In Neoceratodus there is only one lung but in *Protopterus* and *Lepidosiren* there are two lungs. Both aquatic and pulmonary respiration take place in the lung fishes. In Protopterus, only the 4th and 5th branchial arches bear gill filaments though there are six branchial arches. An anterior hemibranch is there on the hyoid arch and a posterior on the 6th arch.

Four pairs of gills are present in *Neoceratodus*. Each branchial arch bearing a double row of gill filaments. The fifth branchial arch is gill–less. In the larvae of Lepidosiren, four pairs of functional gills are present.

(10) In lung fishes the heart is almost three chambered like the Amphibians. The atrium is completely divided into two chambers by interauricular septum but the sinus venosus

is imperfectly divided into two chambers. Blood (oxygenated) returning through the pulmonary vein enters the left side of the atrium, while the deoxygenated blood is poured into the right side of the sinus venous and enters the right auricle. Then by means of the spiral valve within the conus arteriosus, blood from the right auricle changes its direction towards the last (4th to 6th) aortic arches. From here, some of it can get into the lungs, while oxygenated blood from the left auricle is shifted towards the openings of the arteries leading to the head. Thus venous blood goes into the posterior branchial arches and carried to the lungs finally. So, both branchial and pulmonary circulations are met with.

There are generally four functional gills each a holobranch and supplied therefore by aortic arches 3rd to 6th. The afferents are always one per arch in Dipnoi and each holobranch carries two efferents. From near the dorsal end of each 6th arch, there arises a pulmonary artery to the lungs. In larval lungs fishes, capillary loops supply the external gills from the branchial arches 4th to 6th.

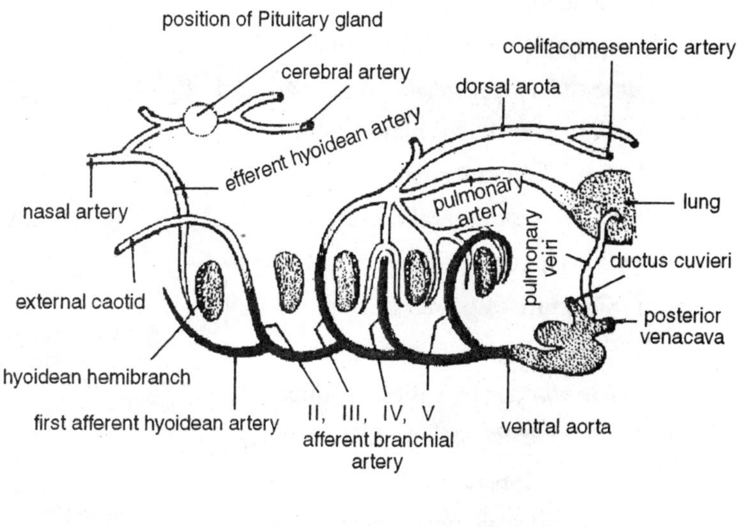

Fig.9 : Blood vascular and respiratory systems in *Protopterus* (Partly based on Parker and Haswell)

(11) One pair of mesonephric kidneys are present. They discharge into the cloaca through the ureters. In the female one pair of ovaries are present. Oviducts open anteriorly into the coelom. They join posteriorly and open into the cloaca. In the male, one pair of testes are present which are close to the kidneys. The process of spermatogenesis is performed by the anterior part of the kidney but the posterior part acts as a vesicula seminalis. The two vasa deferentia join and open into the cloaca. Mullerian ducts are present in the vestigial form.

Affinities of Dipnoi

Characters of Dipnoi show that this group has not undergone much change during the long period of its existence. From time to time they have been linked with several groups of fishes, and amphibians but finally they have been given the status of a separate class because of the presence of several striking features.

A few important points of resemblance with the different groups have been mentioned below :

(A) *Affinity with Elasmobranch*—because of–

(i) Cartilaginous endoskeleton present

(ii) Persistent notochord

(iii) Spiral valve in the intestine.

(B) *Affinity with Holocephali*—on the ground of—

(i) Nature of dental plate

(ii) Autostylic jaw suspension

(iii) Position of Lateral line system

(C) *Affinity with Ganoid fish*—because of—

(i) Structure of paired fins

(ii) Cosmoid scale (similar)

(iii) Air bladder modified as lungs

(D) *Affinity with teleost*—on the basis of—

(i) Similar lobate fins

(ii) Cycloid scale with consmine layer

(iii) Operculum covering the gill opening

(iv) Resemblance with actinopterygii in the structure of paired fins.

(But, for presence of lungs, autostylic jaw suspension, diphycercal caudal fin etc. they were supposed to be distinct).

(E) *Affinity with Amphibians* —

The points of resemblance between the modern dipnoans and amphibians are as follows :

(i) Lungs are present for breating in addition to gills.

(ii) Resemblance of Blood Vascular System as described already.

(iii) Larvae of *Protopterus* and *Lepidosiren* possess external gills like tadpole larva of frog.

(iv) Autostylic jaw suspension

(v) Presence of cloaca in both groups

(vi) Pulmonary artery and pulmonary vein present.

(vii) Large cerebral hemisphere and small cerebellum

(viii) Similar mode of development

(ix) Three-chambered heart in Dipnoi, like amphibians

(x) A pair of external nostrils opening in to the buccal cavity

But, there are some specialised characters also, possessed by the members of this group (Dipnoi) and because of these characters the amphibian affinity has been ignored.

These characters are :

(i) Loss of premaxilla and maxilla

(ii) Lack of ossification in skull parts

(iii) Specailized tooth plates

(iv) Fusion of verterbrae in the back of the skull

(v) Ventral ribs

(vi) Structure of paired fins

Conclusion

Thus the relationship of Dipnoi with fishes and amphibians seems to be deceptive. All the characters possessed by Dipnoi are not present either in fishes or in amphibians. Only a few resemblances are supposed to be due to convergent evolution. Some zoologists held the opinion that Dipnoi are the direct ancestors of amphibians. But this view has been ruled out. Lung fishes seems to by physiologically transitional beetween fishes and amphibians. According to Romer "the lungs fishes are not the ancestors but he uncles of the land dwellers" because the ancestral Dipnoi and crossopterygii are supposed to have been derived from a common stock as conceived by Watson.

There is a general agreement that Dipnoi evolved as a well defined side branch of the bony fishes by the mid–devonian time and were certainly derived from the ancestral *Crossopterygian* stock from which the *Osteolepids* and *Coelacanths* arose. This concept seems to be more convincing. The South American and African lungfishes have diversed as side branches from the ancestral stem of dipnoan evolution. Thus Berg's view of giving the Dipnoi the status of an independent class is plausible.

There is absence of Dipnoi in the oriental region. As we know, the lungfishes do not occur in India at present, but there are evidences that they were present in the geological past. The fossil remains of Dipnoi (genus–*Ceratodus* Agassiz) have been found in the Maleri bedsof the Godavari Valley during the Upper Triassic period, which spread to the Southern hemisphere reaching the area of Assam. Probably during the Upper Cretaceous time they dispersed to South Africa, South America and Australia. Such idea has been proposed by Hora and Menon (1952) and Menon and Prasad (1961).

Holocephali

The Class Holocephali includes three living representatives today and they are *Chimaera, Callorhynchus* and *Harriotta*. All are found in the deep sea oceanic waters. They are the distinct offshoot of the earliest cartilaginous fishes of the Mesozoic era and flourished in the cretaceous. The best known fossil example of the lower carboniferous time is the *Helodus*. Some important extinct representatives of the class Holocephali are : *Squaloraja* and *Myriacanthus*.

Chimaera monstrosa (Chimaeridae) is a very active fish and is commonly known as rat fish or king of herrings. They eat small invertebrates and fishes. They are poor swimmers and move by undulations of the back half of the body. It has an elongated pointed rostrum. They are abundant on the coasts of Japan,

Europe, Australia, New Zealand, North America and Africa. The Pectoral fins are large and fan like, and the tail is elongated into a long whiplash. In the males, there is a club–shaped frontal clasper on the head (Cephalic clasper). In front of the each pelvic fin there is an anterior clasper armed with denticles in addition to ordinary clasper behind the fin. Thus there are five claspers in all. In the mouth there are crushing plates instead of the teeth. A skin flap (operculum) covers the gills on each side. The tail appears to be diphycercal in this fish.

Callorhynchus antarticus in commonly found in the South Pacific. It has a pointed rostrum produced forward and ending in ventrally directed position serving as a tactile organ. In the male, a frontal clasper is also present. Its tail is clearly heterocercal.

Harriotta is found in the North Atlantic. It has a prominent elongated, tapering and depressed rostrum. The pectoral fins are very large. In this the frontal clasper is absent and other claspers are smaller in size. The tail is not turned in the upward direction.

External Features

The body is shark like in appearance. *Chimaera* is about 2 feet long but some fishes reach a length of 6 feet. The mouth is small and it is present in the ventral side. There is a single nasal aperture and four pairs of gills covered by cartilaginous operculum on each side. They open outside by a single branchilal aperture. Eyes are very large. There is absence of spiracle and there is no cloaca. The urinogenital aperture and anus are quite separate and distinct. There are two dorsal fins and a ventral fin. The 2nd dorsal fin in *Chimaera* is much elongated. Pectoral fins are very large but pelvic fin is small. The first has cartilaginous spine. The lateral line is grooved in *Chimaera* but a closed tube in *Callorhynchus*. Skin is generally smooth and slivery. In *Chimaera* the palcoid scales are restricted to certain parts of the body. *Callorhynchus callorhynchus* has a peculiar extended snout, part of which is turned back, below and infront of the mouth. This is a pad supplied with nerves and is sensory in nature.

Endoskeleton : The endoskeleton is cartilaginous. The vertebrae are reduced to separate nodules and the notochord is persistent and unconstricted. The palatoquarate is fused with the neurocranium, providing support for the lower jaw. Mouth has crushing plates, two pairs in the upper jaw and one pair in the lower jaw. They have not enamel but vitrodentine and reduced

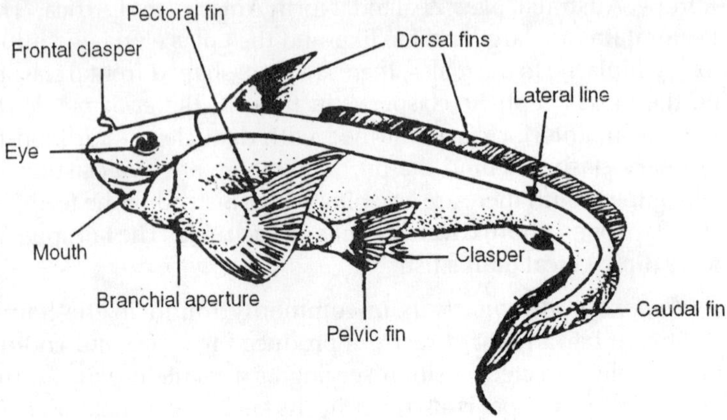

Chimaera monstrosa

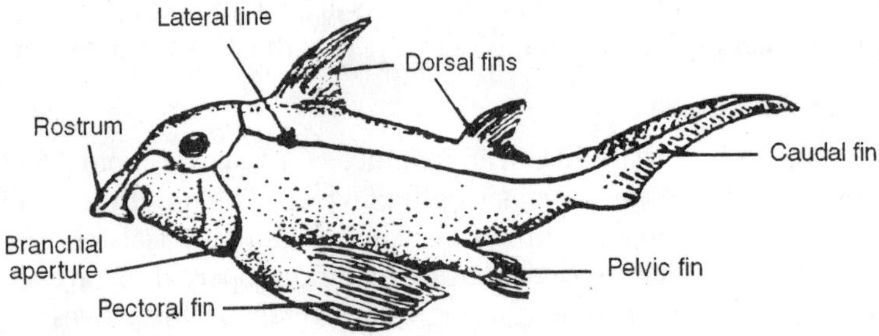

Callorhynchus antarcticus

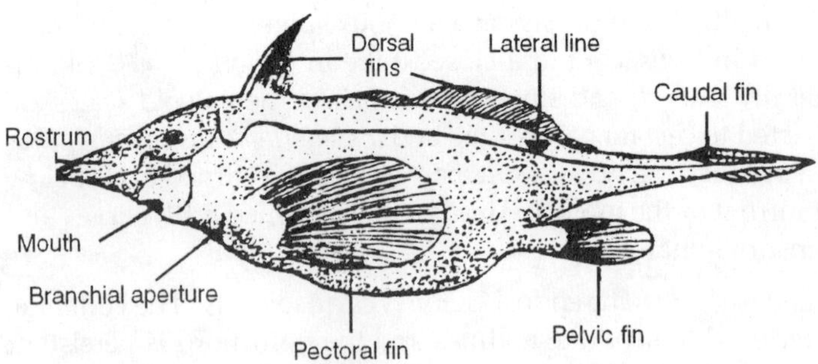

Harriotta

Fig.10 : Members of *Holocephali*

pulp cavities. An inter–orbital septum is present dorsal to the brain case. Two halves of the pelvic girdle are not fused. In the notochordal sheath of *Chimaera* calcified rings are embedded. The neutral arches of the first few vertebrae are fused providing attachment to the first dorsal fin in the form of a plate. In *Callorhynchus,* the snout is supported by three cartilaginous rods which grow from the cranium. These rods are very much reduced in chimaera. The skull is *holostylic.* The Pituitary fossa is comparatively deep and inclined backward. The *condyle* is saddle–shaped articulating with the vertebral column. The hyoid resembles the branchial arches. Above the *epihyal* there is a small cartilage, the *pharyngohyal.* In the first dorsal fin the *pterygiophores* are fused in a single plate.

Digestive Tract : Teeth are present in the form of strong plates with irregular surface and sharp edges.

In the upper jaw there is one pair of *vomerine teeth.* Behind the vomerine teeth there is a pair of large *palatine teeth.* In the lower jaw, there is a single pair of large *mandibular teeth.*

The alimentary canal is straight. There is no true stomach, and there is a special valve in the intenstine. Mouth is adapted to cruch the small invertebrates and fishes with the help of plates.

Nervous system : The cerebral hemispheres are small in size and they are spindle–shaped. The medulla oblongata is laterally produced into large frill–like body which is known as the *restiform bodies.* The diencephalon is long, thin and trough–shaped. The olfactory peduncle bears at its extremity, a compressed olfactory bulb. the Pineal body is small and rounded and has a hollow stalk. The Pituitary body has intra and extra–cranial portions.

Respiratory System : There are three pairs of *holobranchs* (complete gills), two *hemibranchs* (half gills). The fifth branchial arch is gill–less. The holobranchs are present on the first three branchial arches, one hemibranch is present on the first three branchial arches, one hemibranch is present on the anterior face and another on the posterior face of hyoid.

Blood Vascular System : The blood–vascular system of Holocephali resembles that of Scoliodon in all essential respects. The heart consists of *sinus venosus, atrium, ventricle* and *conus arteriosus* like dogfish. *Conus arteiosus* has three rows of valves.

Urino–Genital Organs : The main excretory organ is in the form of a pair of lobed deep–red kidneys which are short and stout in comparison to scoliodon. They are slightly longer in males. The anterior part of the kidney consists of true uriniferous tubules. The kidney is segmented both anteriorly and posteriorly. The posterior part is narrow.

A large number of mesonephric ducts arise from the kidneys and they mostly open into the vas deferens, however, the last six open into the urinogenital sinus. In the female the ducts open into the urinary bladder (urinary sinus).

The female reproductive organs comprise a pair of ovaries, large shell glands and uteri. Reproductive organs are formed on the elasmobranch pattern. In the male there are two ovoid testes containing immature sperms. These sperms pass through the vasa efferentia and reach into the vas deferens which coils to form the epididymis. The epididymis closely applied to the anterior part of the kidney. The sperms aggregate into *spermatophores* surrounded by membranous structure and they are full of gelatinous substance. The bundles of sperms are embedded in the gelatinous material. The lower part of the vas deferens dilates to form the *vesicula seminalis*. This dialated portion is imperfectly divided into compartment by transverse partitions. The spermatophores are passed into these compartments and then pass into the urinogenital sinus. The fertilization is internal and the egg becomes surrounded by a horny egg–shell secreted by the shell–glands.

Affinities

The Holocephali show some resemblances with the elasmobranchs and some with the teleosts. Besides these characters, this group possesses a number of specialized characters which are not present either in Elasmobranch or in teleosts and as such, this group has been given the status of independent class. The characters have been analysed below :

Resemblances with the Elasmobranchs

(1) In Holocephali the skeleton is cartilaginous, replacing and membrance bones are absent.

(2) The vertebral column has a persistent notochord with cartilaginous arches.

(3) Limbs and girdles are formed on the elasmobranch plan.

(4) The skin is silvery and smooth but some placoid scales are present on the claspers.

(5) A pair of claspers are present posterior to the pelvic fins in the male (like elasmobranch) however, anterior and frontal claspers are also present.

(6) The basic plan of the brain resembles with the elasmobranch but the development of restiform bodies and elongation of diencephalon are remarkable differences.

(7) A spiral valve is present in the intestine.

(8) The air bladder is absent like elasmobranchs.

(9) The tail is heterocercal in many holocephalians.

(10) The Urinogenital organs are of the selachian type.

(11) The conus arteriosus is present in the heart and contains three rows of valves as in the case of elasmobranchs.

Resemblances with Teleosts

(1) Gills do not open outside, rather they are covered by means of operculum like teleosts.

(2) There is a single external branchial aperture.

(3) The gill filaments project beyond the inter branchial septum, the latter being much reduced.

(4) The spiracle is absent.

(5) There is no cloaca and the urinogenital aperture lines behind the anus.

Special Features of Holocephali

In addition to the above mentioned characters, the Holocephali have some specialized characters also which are as follows :

(1) The skull is of holostylic type and the palatoquadrate is fused with the cranium.

(2) Extra clasper (Anterior and Cephalic claspers) are also present unlike other groups.

(3) The dention is peculiar in having large plates. Teeth are not covered by enamel but a layer of vasodentine is present. Sharp edges are there in the plates. The pulp cavity is much reduced.

(4) There is loss of scales and spiracle.

Conclusion

This in Holocephali we find some characters like elasmobranchs and others like teleosts. Besides, there are some special features also, already mentioned. The *Holocephali* arose from some shark ancestor during the Mesozoic era, possibly in Triassic period. Originally the Holocephali were considered as an order under the subclass–*Bradiodonti* of the Class—Elasmobranchii but now it has been given the status of independent class and it seems to be justified.

Accessory Respiratory Organs in Fishes

Fishes are able to breathe oxygen dissolved in water by means of gills which are richly supplied with blood capillaries. Exchange

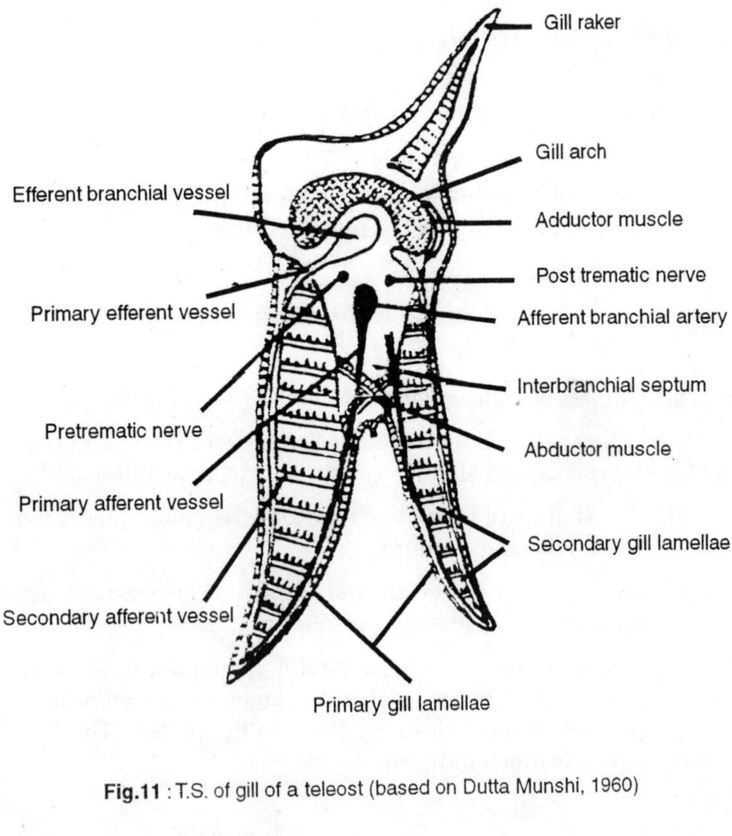

Fig.11 : T.S. of gill of a teleost (based on Dutta Munshi, 1960)

of oxygen and carbondioxide takes place mainly through gills. But, sometimes their aquatic gills are insufficeint in taking required amount of oxygen and getting rid of carbondioxide. So they have to take oxygen from air and as such, the fishes have evolved many respiratory adaptations for air–breathing. Thus, there are certain modified structures which assist the function of gills. These organs are known as accessory respiratory organs. Thus, accessory respiratory organs are those organs that assist in the process of respiration. There are several fishes which have air–breathing tissues and are capable of spending several hours out of water. Air breathing function can be performed not only by the pre–existing structures like intestine, skin, gills, air bladder etc., but certain newly formed structures are also capable of utilizing atmospheric oxygen. These new structures are known as *neo–morphic* air breathing organs. They develop respiratory epithelium for this purpose.

Some notable contributions on the accessory respiratory organs of fishes are those of Das (1927, 1940), Sawaya (1946), George (1953), Munshi (1961, 1962, 1968), Johansen and Hanson (1968), Johansen and Lenfant (1968), Johanesn (1968, 1970), Johansen *et. al.* (1970), Prasad and Mishra (1982), Singh *et. al.* (1990), Munshi and Hughes (1992), and others. The following are the important accessory respiratory organs in fishes:

1. *Wet Skin Surface* : Any thin walled tissues, richly supplied with blood capillaries and having moist surface can easily serve as the organ for respiration, when it comes in contact with oxygen. In many fishes the skin is richly vascular and these fishes can respire through their skin which must be kept moist. *Anguilla anguilla* and *Amphipnous cuchia* are very common examples. When these fishes leave water, they can pass through damp vegetation and cover a short distance, thus making a land journey. Some larval fishes breathe by skin until the gills are fully formed. Cutaneous respiration is performed also by Mudskipper (*Periophthalmus*).

2. *Parts of Alimentary Canal* : In certain fishes, various parts of the alimentary canal function as respiratory organ. The method has also been adopted by loaches (cobitidae) and some catfishes (Loricariidae). The pond–loaches (*Misgurnus fossilis*) swallows air and its intestinal wall extracts oxygen. They use the intestine at times for

respiration. In the loaches, the middle and posterior parts of the intestine function both as digestive and respiratory organs. In some of the loaches the digestive phase of the intestine alternates with respiratory phase. In certain cases the intestine serves as the respiratory organ during the summer months only. The weather fish of Europe swallows air and the latter is passed down into the intestine. The vascular reservoir of the intestine extracts

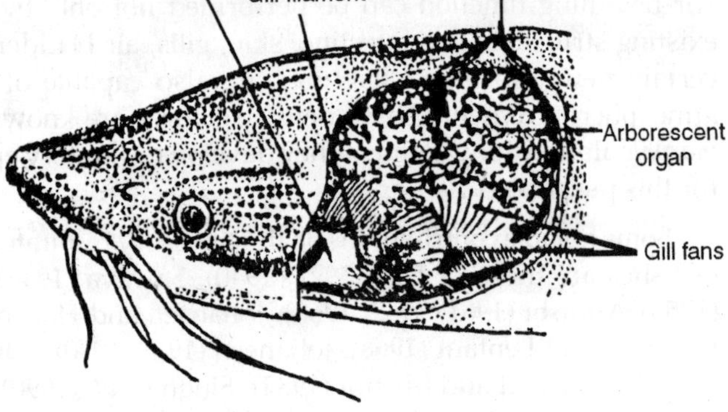

Suprabranchial organ of *Clarias mossambicus*

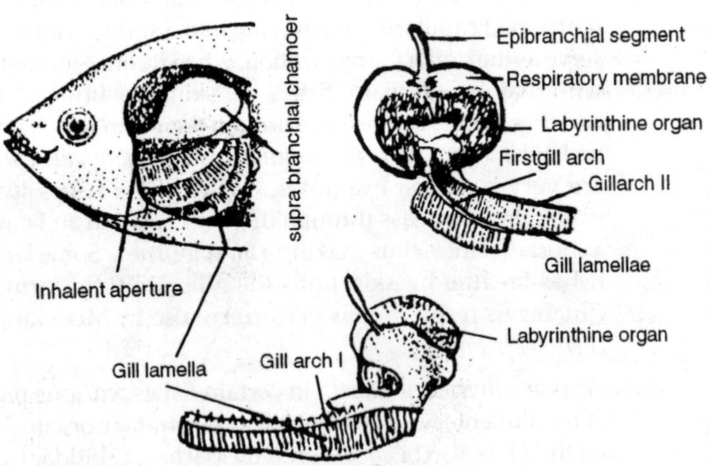

Trichogaster faciatus (after Datta Munshi)

Fig.12 : Accessory respiratory organs in two fishes

oxygen from the air. The remaining gases are then emptied out through the vent. The rectum of the nesting catfish (*collichthys*) absorbs oxygen drawn in through the anus and hind gut i.e rectum absorbs oxygen from that.

3. *Arborescent Organs* : Gill cavity enlargement is found in *Clarias* (clariidae) and related catfishes. Opening between the second and third gill arches leads to two sacs with bush like highly vascular structure known as arborescent organ. They are tufts of epithelial outgrowths and has cartilaginous supports which originate from II and IV gills. The entire structure inside the sac received its blood supply from all the afferent branchial arteries and oxygenated blood returns to the corresponding efferent vessels. The air–breathing organs of *Clarias batrachus* comprises (i) Suprabranchial chamber, (ii) Two beautiful rosettes of air trees, (iii) the fans, and (iv) respiratory membrane.

   The suprabranchial chamber lying above the gills is divided into two cup like compartments lined by a vascular membrane. Two beautiful "rosettes" (dendritic organs) are present on each side. These are supported by the epibranchial cartilages of the II and IV branchial arches. The rosette consists of a core of cartilage and is covered by vascular epithelium showing eight folds. Fans are formed as a result of the fusion of some gill lamellae of each arch. The vascular area of the supra–branchial chamber shows several "islets." The gill–slit between the second and third gill arches serves as the inhalent aperture, through which air enters into the supra–branchial chamber. Air entering into the opercular cavity is directed into the suprabranchial chamber by two fans present on the 2nd and 3rd gill arches. The gill slit between the 3rd and the 4th gill arches functions as the exhalent aperture through which the air from the suprabranchial chamber goes into the opercular chamber and finally to the exterior. Partial vacuum in the supra–branchial chamber is created by its contraction and also by movement of the fans. When mouth opens, the buccopharygeal cavity is enlarged and the air is inhaled.

4. *Saccular Organ* : In *Amphipnous cuchia* (synbranchiformes) the air–breathing organs consist of a pair of sacs which

grow out from the pharynx just above the gills and air can be stored in the sacs for sometime, for gaseous exchange. The gills in this fish are rudimentary and the power of aquatic respiration has been lost. It comes to the surface of water to gulp air at frequent intervals. Each sac is provided with inhalent and exhalent apertures. The epithelium of the air sacs consists of vascular areas in the form of "Islets" having numerous "rosettes" of papillae. *Saccular organ* is also present in *Heteropneustes fossilis* (Bloch). Here the sac is very long and reaches the tail region. The gills are fused to form a valve over the orifice of the air sacs. The IV branchial artery supplies blood to this structure and the V efferent drains the blood. The "islets" are also covered by a thin epithelium which has numerous capillaries.

5. *Labyrinthiform Organ* : This organ is present in the climbing perch, *Anabas,* (Anabantidae) and has been thoroughly studied. This organ is derived from the epibranchial segment of first gill arch and is covered with fold of respiratory epithelium. The *Labyrinthifrom organ* is located in the pocket of the pre–opercular dorsolateral area of the head region. This organ is also present in *Betta* and *Trichogaster*. The air–chamber in *Anabas testudineus* communicates with the buccopharyngeal cavity and opercular cavity. There is lining of the vascular epithelium in the air chamber. The epithelium is sometimes folded. The labyrinthine organ is lodged in this air chamber.

   In *Trichogaster* there is a supra–branchial chamber situated above the gills on either side communicating with the phyarynx by inhalent aperture and with the opercular chamber by means of exhalent aperture. The labyrinthine organ is spiral in form with two leaf like expansions covered by vascular epithelium. The vascular area of the labyrinthine organ comprises large number of "islets" with parallel blood capillaries. The islets are derived from the secondary lamellae of a gill filament.

6. *Accessory Branchial Chamber* : The accessory branchial chamber is present in *Channa* sp. Because this chamber lies above the usual respiratory chamber, it is also known as supra–branchial chamber. There are two chambers which are incompletely separated. Diverticula of the

mouth and pharyngeal cavities develop. The diverticula are provided with folded respiratory epithelium which has a rich blood supply from the afferent vessels. In *Channa marulius* and *Channa striatus* the suprabranchial cavities are developed in the roof of the buccopharynx. These cavities are highly vascular. It is lined with respiratory epithelium which is raised into folds and tongues and shows respiratory "islets". The branchial cavities possess some alveoli also. The flat epibranchial of the first gill arch is also covered by vascular membrane which assists respiration. It also has a highly branched respiratory labyrinthine organ.

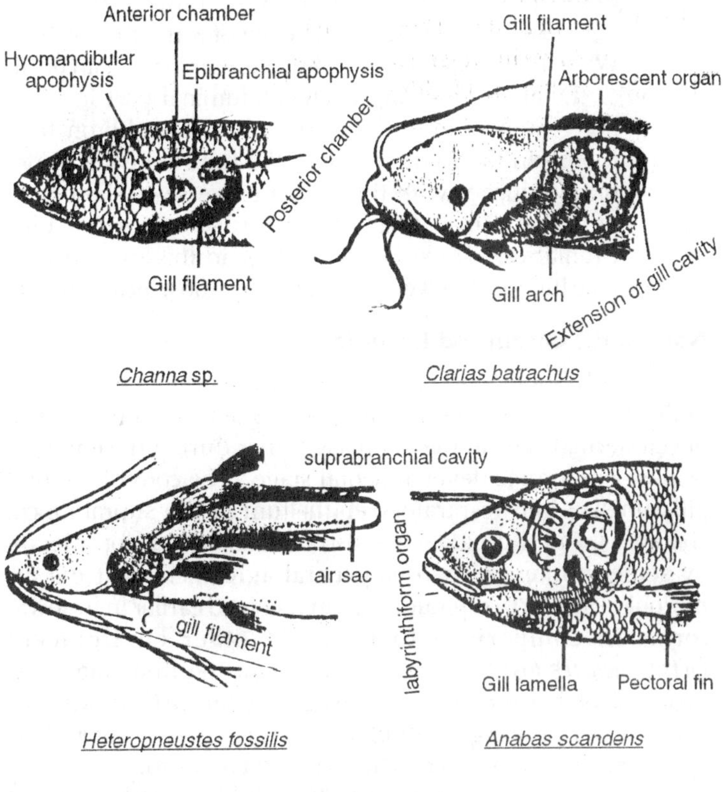

Fig.13 : Accessory respiratory organs in some more fishes

7. *Air Bladder* : Many fishes use the air bladder as a temporary or supplementary organ of respiration. The gars (*Lepisosteus*) and the bow fin (*Amia*) are examples of facultative air–breathers with a highly sacculated or alveolar air–bladder. In lung fishes, the inner walls have become thin and richly vascular. It is partitioned by the development of Septa and cross septa to form a series of alveoli. Thus the surface area is greatly increased. *Neoceratodus* of Australia comes to the surface of breathing air. In *Protopterus* of Africa and *Lepidosiren* of South America, each lung sac cavity is further divided into a series of large and small irregular alveoli. The alveoli sub–divide into tubules which finally terminate into small pockets. In *Notopterus chitala*, a physostomous fish, the gills are reduced and the air–bladder is very much developed and is almost like the lung of the higher vertebrates. The fish can leave out of water for few hours. When in water, it periodically gulps air. The main portion of the air bladder lies in the abdominal part inside the body cavity and bifurcates into two lateral bronchi like tubes. In the trunk region of the tube there are 14 or 15 blind pouches with finger like processes. The pouches and their processes have profuse supply of blood capillaries. Thus the air bladder in this case serves as true lung and makes the chital fish almost an air–breather.

Note on the Origin and Evolution

Recently some workers have based the origin and evolution of the neo–morphic air–breathing organs of teleosts on induction of ectodermal cells in the gill mass, either during the embryonic or post–embryonic developmental stage. The conversion of the gill lamellae into respiratory epithelium of the suprabranchial chamber (as in siluroidae) has also been taken into consideration by some workers. There is general agreement that the neo–morphic accessory respiratory organs evolved earlier in freshwater forms in comparison to fishes of other environments. *Heteropneustes fossilis* is said to represent the first stage in the evolution of these organs in freshwater habitat. In *Channa morulius* (Channidae) these organs are of the most archaic type but in *Channa gachua* it is quite advanced in the evolutionary history. *Monopterus cuchia* as revealed by Singh *et al.* (1990) is more evolved than *M. albus*. In the family Anabantidae, *Osphronemus nobilis* is the most primitive but *Anabs testudineus* is the most evolved when critically examined. Prasad *et.al.* (1982) have revealed that *Colisa fasciatus* occupies the intermediate position between the above two forms.

Parental Care

Rearing or care of the offspring is a very important achievement in the trend of evolution. This phenomenon of parental care is well developed in several groups of fishes.

There are various ways and means which facilitate the fishes in overcoming the hostile forces. It is actually very important in perpetuation of the race.

Fishes are lowly creatures and produce milllions of eggs every year. There are large number of interesting cases where parents protect and take care of the fertilized eggs and the young ones. Various means for affording care to fertilized eggs and the young ones by one or both sexes among fishes are remarkable. Selection of spawning sites, laying of eggs at suitable places, nest building, protection of eggs and young ones are some of the important aspects of parental care.

Some *anadromous* fishes (*Acipenser, Salmo, Onchorhynchus*) ascend freshwater streams for spawning. The freshwater eels (Anguilla) have a *catadromous* habit, descending into the ocean for laying eggs.

Some fishes are efficient nest builders. The nest may be a simple trough or a hollow cleaned out in the sandy bottom of the stream where the eggs can be laid conveniently. In this case the male guards the eggs. The *Salmon* is a common example in this case.

The mud fish (*Protopterus*) hollows out in the mud of the swamp rich in aquatic weeds and grass which afford protection. This hollow becomes the nest in which the female lays eggs. The nest is built mainly by the male. It is the male who takes care of the eggs. He swims around the nest and chases the predators. Sometimes he aerates the water of the nest for the benefit of the eggs and young larvae.

The African Osteoglossid (*Clupisudis*) makes the nest by cleaning a space in the aquatic vegetation. The nest is built in about 2' of water it is about 4' across. The walls of the nest is several inches in thickness, made up of the stems of grasses. The floor of the nest is constituted by the smooth bare ground.

Tilapia mossambica (female) providing shelter to young ones (in her buccal cavity)

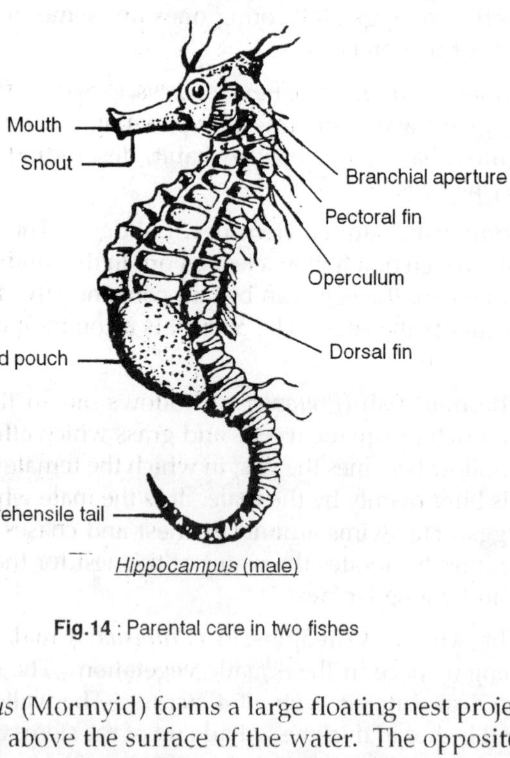

Hippocampus (male)

Fig.14 : Parental care in two fishes

Gymnarchus (Mormyid) forms a large floating nest projecting several inches above the surface of the water. The opposite end forms an entrance about 6″ below the water.

The nest of *Gasterosteus* is much more elaborate. The male labours hard for gluing fragments of plants together. The glue is a special product of the kidneys. When the nest is ready, the male goes in search of a female who comes and lays some eggs in the nest. Some of the eggs are fertilised by the male. The male again goes in serach of another female, courts her also and invites her to the nest. The process is repeated till the nest is completely filled up by the eggs. The male now guards the eggs till they hatch out and the young ones leave the nest subsequently.

North American *Amia* also constructs a crude nest which is circular in appearance. The nest is placed at the swampy end of the lake rich in aquatic herbage. When the nest is completed one or more females attend. The fertilized eggs adhere to the bottom materials of the hollow. Eggs are guarded by the male. When the young fishes hatch, they leave the nest in a body under the care of the male who keeps them together by encircling around them.

Some catfishes (Amiruidae) of North America make a crude nest in the mud. In this process of excavation both parents share. Sometimes the nest is also placed in crevices in the river banks, beneath logs, under the stones or even in buckets or other receptacles in the water.

Wrasses (Labridae) can also construct crude nests of sea-weeds, shell or stones and in this process both sexes share. The fighting fish '*Betta*' makes a nest of foam. It blows a surface bubble nest. The male makes bubbles of air and then glues them by mucous from his mouth. Thus a dome–shaped floating mass is formed. After elaborate courtship, the females lay eggs whch are fertilised by the male. The male catches the eggs in his mouth and after providing a coating of mucous, sticks them to the undesirable of the nest. Finally, about 150 to 200 eggs are massed together in the nest. Both parents guard the nest till the larvae hatch and young ones leave the nest.

The Butter fish (*pholis*) lays its eggs in the valves of empty oyster shells. Sometimes the eggs are also laid in the holes of rocks made by the boring molluscs. The mass of spawn is just like a ball, about the size of a Brazil nut. Both male and female fish cooperate in this operation. Now, whether the male guards the eggs by coiling around them or the female does so, is not clearly known. It is probatale that both parents take their turn in guarding the eggs.

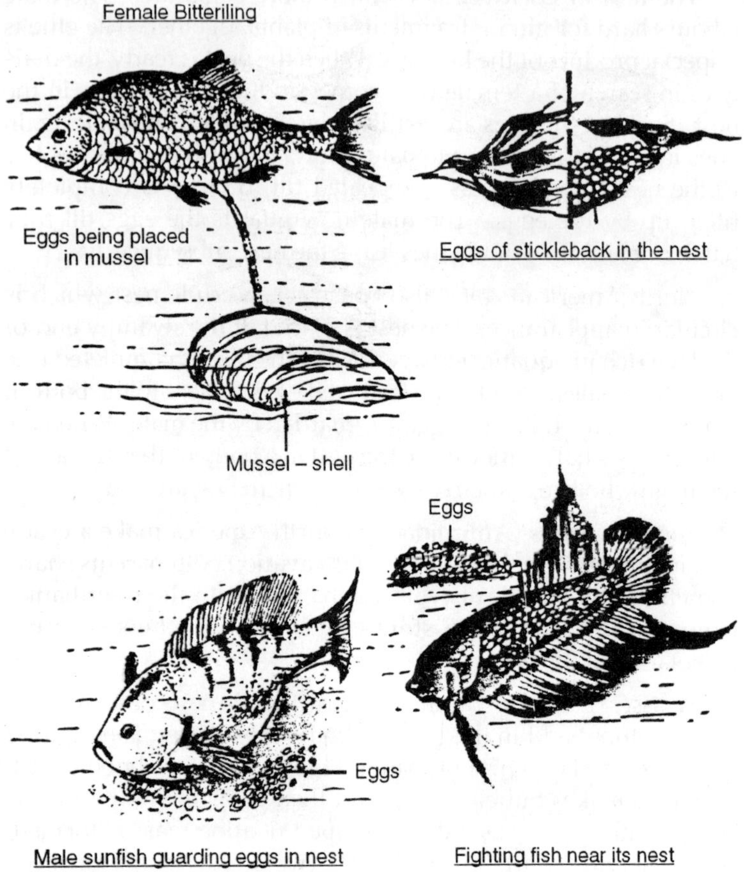

Fig.15 : Parental care in some more fishes

In the genus *Kurtus* (Perciformes), the male has a bony hook projecting from the fore-head. The egg mass becomes attached to this hook in such a way that one bunch of eggs lies on either side of the male when he swims in water.

The eggs of the *Rhodeus* (European carp) are deposited in the mantle cavity of the freshwater mussel. The eggs undergo development inside the shell. The eggs are also aerated by the respiratory current of the mussel.

In the Brazilian catfish (*Platystacus*) the lower surface of the body of the female becomes soft and spongy when the skin swells

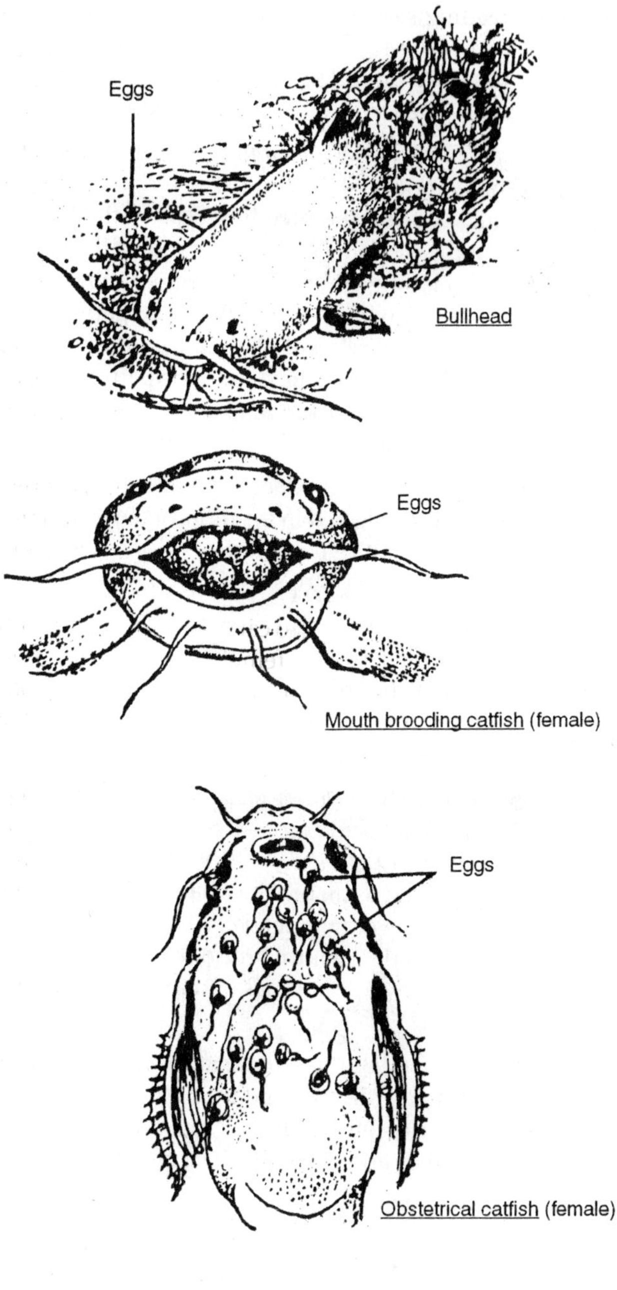

Fig.16 : Some examples of Parental care

during the breeding season. The female lies on the eggs when they are fertilised. The eggs become attached to the skin by small stalked cup–like structure. The cup is vascular and it nourishes the embryo. Thus the fertilised eggs are fixed until hatched.

In the *Hippocampus* and *Syngnathus*, the eggs are placed by the female in the brood pouch of the male. The brood pouch is present beneath the tail. During the breeding season the brood pouch becomes thick and vascular. The vascular wall of the brood pouch nourishes the developing embryos. The eggs are transferred into the male brood pouch with the help of genital papilla developed in the cloaca of the female. It acts as an intromittant organ. When the embryo is completely developed, certain apertures in the pouch become visible and through these apertures, miniature sea horses come out.

In the marine catfishes (Aridae) the male undertakes the care of eggs. The eggs are collected in his mouth. They are retained in the mouth during the entire period of development. During this period the male completely stops feeding.

The African cichlid fishes also carry the eggs in the mouth and this duty is performed by the female in this case. Even after hatching the young fishes do not leave the shelter of the maternal mouth so easily. Later on, they swim in about in small shoal, accompanied by their parents.

In the Brazilain fish, *Loricaria typus*, the lower lip of the male enlarges to form a pouch and becomes suitable for labial incubation of the eggs. In two catfishes, *Bagre marinus* and *Galeichthys felis*, the mouth is employed as an oral incubator. Each male carries in his mouth 10 to 30 developing eggs. The African fishes, *Tilapia mossambica* are called "mouth brooders" because the newly hatched young ones escape at times of danger into the oral cavity of the female.

Recently, Coyer (1982) has reported that in giant kelp-fish, *Heterostichus rostratus*, the male guards the eggs after the spent females leaves. Male defence of the eggs–laden nest is vigorous against all cospecifics and some algae searching fish predators.

Chapter 4

Class – Amphibia

The animals of Class – Amphibia (Gr. Amphi = Both, bios = life) live both in water as well as on land, both habitats being essential for them. They are cold–blooded (Poikilothermal or ectothermal) animals who appeared as first tetrapod during the devonian period, possibly from the *crossopterygian ancestor*. They have two pairs of pentadactyle limbs supported by girdles. The skin is always soft, granular and moist. Normally, two nostrils are connected with mouth cavity. There are two occipital condyles in the skull. They respire by gills, lungs, skin and epithelial lining of the buccal cavity. Winter sleep (hibernation) and summer sleep (aestivation), to avoid excessive cold and heat respectively, are very common among most of the amphibians. Fertilization is either external or internal and they are oviparous. There are 10 pairs of cranial nerves in the animals of this class. The development is indirect, a tadpole larva appears which later on metamorphoses into the adult form.

Origin and Evolution

Introduction

Amphibia has been considered as a transitional group which has a made a sort of compromise between the aquatic life of fishes and terrestrial life of the tetrapods. It has already been pointed out that both aquatic and terrestrial adaptations have become essential within the range of amphibian organization. Complete terrestrial adaptation has become practically impossible. Devices have been developed to meet the necessities of aquatic and terrestrial habitats. Class – Amphibia has been placed between Pisces and Reptilia. The origin of tetrapods is essentially the problem of the origin of Amphibia. It has been suggested that the

Devonian time during which the land adaptations originated, was certainly a time of seasonal drought when life in freshwater would have become difficult. Amphibians lead their early life in water. The changes which occur during metamorphosis carry great significance in the light of the Haeckel's *recapitulation* theory.

Various characters of amphibians have been analysed below to trace the relationship of this group to find out an axiom regarding their origin and evolution.

Piscine Origin

(1) *Evidences from Embryology* : Most amphibians spend their early life in water. The tadpole larva of frog has some of the very important piscine characters e.g. presence of fin, presence of gills and presence of lateral line system. Haeckel's recapitulation theory illustrates that "ontogeny repeats phylogeny" and on the basis of this theory, there remains little doubt in confirming that the Amphibians have descended from piscine ancestors.

(2) *Evidences from Palaeontology* : The fossil fauna of Palaeozoic era provide evidences of the ancestry of amphibians. Fossil records of Labyrinthodontia, Lepospondyli and Phyllospondyli have been obtained from *Carboniferous* to *triassic* periods. This indicates that amphibians must have originated earlier in the *Devonian* and possibly in the *Silurian* times. During the Devonian period, foot prints of amphibians have been reported. Amphibian fossils were procured from the freshwater deposits. So, it seems possible that the amphibians arose from Devonian piscine ancestors. It also reveals that the fossils of Devonian time were the sharks, Dipnoans and the early Ganoids (Crossopterygians).

(3) *Evidences from Morphology* : As there are several similarities between fishes and amphibians, Huxley put them together into *Ichthyopsida* which was quite separate from other tetrapods. Points of similarities are :

(i) Cold blooded condition

(ii) Presence of gills and gill–slits (particularly in the larval stage)

(iii) Functional pronephrous during development but mesonephrous kidney in the adult condition.

(iv) Similarity in cranial roofing, palate and lower jaw elements

(v) Lateral line system

(vi) Presence of cloaca

(vii) 10 pairs of cranial nerves

(viii) Larval life in water and also sometimes permanent aquatic life is seen in Amphibians

(ix) Urinogenital system.

Thus, amphibians possess several piscine characters. It is not plausible that the amphibians evolved from the modern fishes. The ancestors of amphibians must be either dipnoi or the early ganoid like crossopterygians. Piscine origin has been supported by Gregoy (1919), Watson (1926), Williston (1925), Romer (1945) and others.

Now, to trace and ascertain the actual ancestor of amphibians, three possibilities are to be critically analysed and they have been discussed below :

(A) *Shark Origin* : On the basis of fossil records, it has been brought into light that the amphibian ancestry is a history of freshwater evolution and as such, the sharks being marine, can not be the ancestor of amphibians.

(B) *Dipnoan Origin* : Of course, there are some resemblances between Dipnoi and modern amphibians, e.g.

(a) both breathe by lungs

(b) skull, brain, urinogenital system and development show resemblances

(c) circulatory system also similar.

But, these resemblances are not very significant. That the Dipnoi can not be the direct ancestors of amphibians, has been claimed on the ground that the Dipnoan fossils are also found in the Devonian time alongwith the amphibian fossils. Thus, the Dipnoan ancestry has been disapproved.

(C) *Ganoid Origin* : Naturally, the only choice now lies with the ganoids which can be considered as the ancestor of 1st tetrapods. One family of Devonian *Crossopterygian ganoids*, the *Osteolepidae* resembles much with the primitive amphibians in all the skeletal features. There is general agreement that first tetrapods branched off from a fish very closely allied to this family.

So far as the beginning of tetrapod organisation in fishes is concerned, neither the dipnoans nor the crossopterygians are considered, but, the osteolepids are to be taken into account. The *Embolomeri* an order under super order – Labyrinthodontia comprises the most primitive amphibians which lived from lower carboniferous to permian periods.

Crossopterygians, particularly oteolepidae fulfill all the requirement for tetrapod ancestor. They are close to *Embolomerous amphibians*. The skull roof patterns of osteolepids and embolomerous amphibians are similar. Lateral line canals crossed the same bones. Spiracular notch of osteolepid was identical to otic notch of embolomeri. Palate and Jaw structures were comparable. Internal nares were present in both, similar lung respiration. Though the paired fins of crossopterygians were smaller in size however, the basic structure was almost similar to tetrapods.

Conclusion

Now, there is a general agreement that the land vertebrates (1st tetrapod) sprang from the *Crossopterygian stock*. Hence the Noble's suggestion that the ancestral forms of amphibia are not the modern Dipnoans nor the crossopterygians nor the fossil dipnoans but the osteolepid–crossopterygians of the Devonian time. It seems to be justified. Osteolepids really appear to be close to possible tetrapod ancestors. There has been evolution and adaptive radiation in modern amphibians. It has been suggested that Labyrinthodonts gave rise to *anurans*. The *Urodels* are supposed to be derived from certain microsaurians (*Lepospondyli*) and the gymnophiones possibly evolved from *Adelospondyli*.

Additional Relevant Informations

Advancement of Amphibians over fishes : For leading a terrestinal life, in addition to aquatic life, certain changes took place. The changes undergone are —

(1) Loss of gills, development of lungs and nasal passage extended to form internal nares.

(2) In many cases, exo–skeleton was lost and to keep the skin moist, skin glands have developed.

(3) For locomotion, tetrapod type limbs developed as the modification of Lobe fins of crossopterygians.

(4) Lateral line system was no longer required, much development of auditory, optic and olfactory organs took place.

(5) The loosely hung jaws became firmly fixed to the brain case.

Era	*Periods*	*Fossil records*	
Palaeozoic	Permian		
	Carboniferous	Phyllospondyli, Lepospondyli, Labyrinthodontia	
	Devonian	Foot prints of amphibians	
	Silurian	Ganoids Dipnoans Elasmobranchs	} To trace which one is the ancestor of amphibians

Parental Care in Amphibia

Like some other groups of animals, in Amphibia also, the parents protect and take care of the fertilized eggs and the young ones. The phenomenon of parental care is well developed in Amphibians. It is actually very important in perpetuation of the race. Various means of affording care to fertilized eggs and young ones by the parents have been described below (in the orders – *Anura, Urodela* (*caudata*) and *Gymnophiona* (*Apoda*).

Order — Anura

1. *Indirect nursing — Protection by means of Nests and Nurseries*

(a) *In enclosures in the water :* A tree frog (e.g. *Hyla faber*) builds a basin–shaped nursery in shallow water on the border of the pond. Normally, the female hollows out the mud to a depth of about 3–4 inches with the help of webbed feet and thus a nest is constructed. The eggs are laid in this nest and the eggs and larvae are protected.

(b) *In holes near water :* On the edge of the ditch or flooded rice field a hole or chamber is made by the genus *Rhacophorus*. The female produces a secretion from cloaca. The male uses its hind feet to whip up a froth from the secretion or from the albumen exuded with the eggs. The nest is thus formed and it is suspended in vegetation over the water. When the tadpole larvae have hatched, the nest liquefies.

(c) *In Nests on trees or on rocks overhanging water : Phyllomedusa* (of South America) makes nest in foliage above water. The females lay their eggs between two leaves which they stick together. The eggs can be laid even in a single folded leaf.

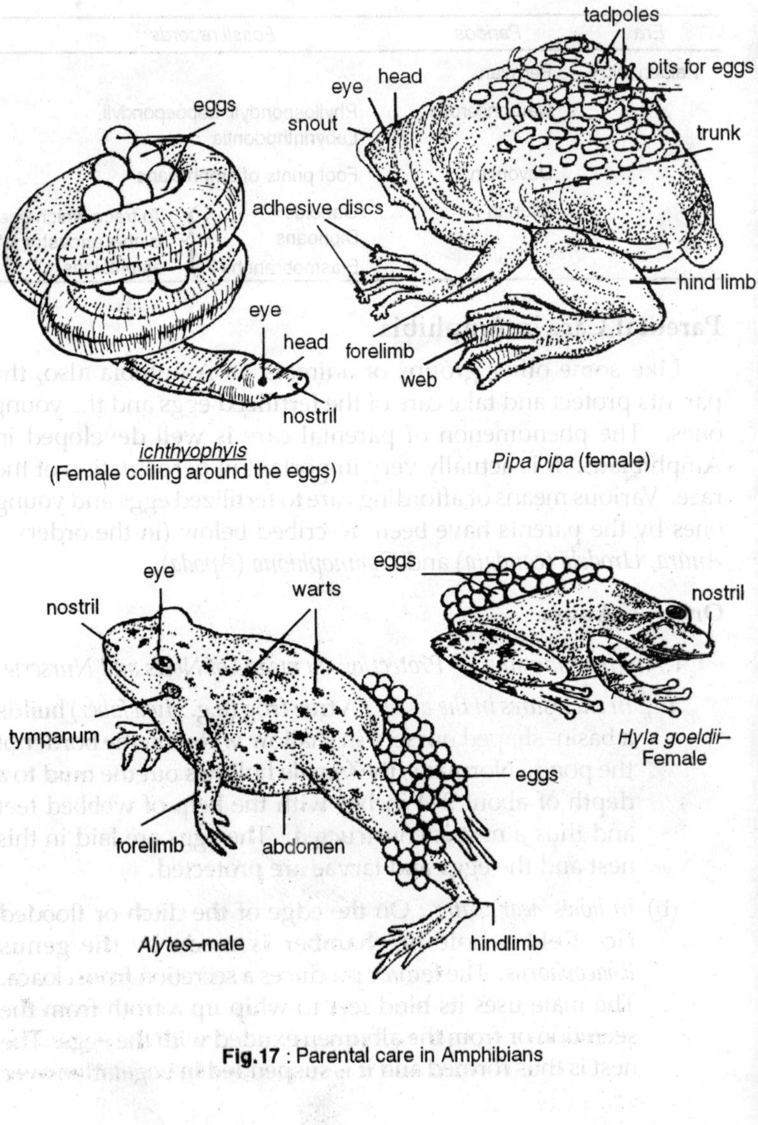

Fig.17 : Parental care in Amphibians

The larvae move in froth and when the external gills are lost they fall in water where the metamorphosis completed.

(d) *In transparent gelatinous bag in the Water*: The eggs of genus *Phrymixalus biroi* (of New Guinea) are laid in a sausage shaped shallow transparent membranous bag which is secreted by the female and is left in the streams. The development takes place in the bag. The miniature frogs leave the bag finally. The large tail serves as the organ of respiration as the gills are absent.

(e) *On trees and weeds near water* : In *Hylodes* (of America) the eggs are deposited in damp places, under stones and leaves of plants near water. There is sufficient yolk in the eggs and as such, entire development takes place in the eggs. Metamorphosis takes place quickly within the eggs. The young frog comes out as an air breather. No gills have been reported.

2. *Direct Nursing by Parents*

(a) *Transport of Tadpoles from one place to another* : The south American frogs *Phyllobates* and *Dendrobates* carry fully developed tadpoles on their back. The sucker–like lips and flat abdomen facilitate and tadpoles in adhering to the back of the parent. Thus they can be transported from one place to another when needed.

(b) *Eggs protected by the male*: In the case of *Mantophryne robusta* the male covers the eggs with its body. Eggs laid by the female are strung together by an elastic, gelatinous envelope. The male holds the eggs with both hands and development takes place as usual. The larvae have no gills. The large tail is vascular and serves as respiratory organ also.

(c) *Eggs carried by parents*

(i) The male Midwife Toad (*Alytes obstetricans*) takes care of the fertilised eggs. The eggs are tangled round the thighs and full care is taken till the eggs hatch.

(ii) In the genus *Gastrotheca*, the eggs are packed by the male into a pouch on the back of the female. From here the tadpoles or baby froglets emerge later on.

The Brazilian tree frog *Hyla goeldii* and British genus *Hyla enansi* take charge of the eggs on their back.

(iii) In *Pipa americana* about 100 eggs are laid and they are carried by the female on its back where the skin thickens and grows round the eggs. A cell like pouch is formed. The cell is covered by a lid probably formed by the secretion of skin glands. Development takes place in the pouches and the young ones come out in a perfect condition.

(iv) In the genus *Rhinoderma darwinii* the male swallows the eggs which develop in its vocal sacs. The modified vocal sac is known as gular pouch.

(v) In *Nototrema marsupium*, a common large pouch develops on the back of the female during the breeding season. The horse–shoe shaped fold of skin in the sacral region gradually deepens and becomes almost a sac like structure. The eggs are filled in this pouch where the development takes place. The young ones escape through the aperture of the pouch in the forms of tadpoles.

(vi) In the case of *Rhacophorus reticulatus* (of Ceylon) the female carries eggs on its belly. Shallow impressions can be seen when the eggs are removed.

(vii) The female of west African *Hylambates breviceps* carries the eggs in her mouth. The eggs are larger in size but very few in number.

(d) *Viviparity* : In case of two East African toads, *Pseudophryne vivipara* and *Nectophryne tornieii* have been reported to be viviparous. The young ones develop with in the oviduct (uterus) of the female. Fertilization of the eggs in the female occurs after the male sperms are directly emitted into her cloaca.

Order – Urodela (Caudata)

Protection by means of nests or nurseries

(a) About 10-20 eggs are laid by *Autodax* in a dry hole in the ground or in a hole on a tree. Both parents remain in the hole to provide moisture and suitable temperature to the developing embryo. The young ones also remain in the hole for some time.

(b) *Salamandrella keyserlingii* deposits all eggs in a gelatinous bag which is transparent. This bag is attached with an aquatic plant (near water). The bag can contain about 60 small eggs. When larvae are hatched, they come into the water.

Direct Nursing by Parents

(i) In *Plethodon* the eggs are laid in packages beneath the stones and the mother coils around them when the gills are lost, the larvae come out.

(ii) In *Megalobranchus maximus*, the male coils round the eggs which are connected by a string. The larvae have three pairs of gills when hatched.

(iii) In *Desmognathus fusca*, the eggs are laid in the form of rosary–like strings which are bound round the body of the female and the latter nourishes them at dry spots.

(iv) *Viviparity* : Copulation in Salamanders (e.g. *Salamandra maculosa*) takes place on land and after several months the female goes to water site and gives birth to 10-50 young ones of small size with fore limbs developed. In the case of *Salamandra atra*, the young are retained in the uterus till the metamorphosis is completed.

Order – Gymnophiona (Apoda)

In *Ichthyophis glutinosa* the female makes a hole in the damp ground near a water body and deposits the eggs in the hole. The mother coils around the eggs with its snake like body. The eggs are 8-10 mm. in diameter and they form a bunch. The gills are lost before the embryo hatches. Some members of the order – Adopa are viviparous also (e.g. *Dermophis*).

Gymnophiona (Apoda)

The Gymnophiona or worm like caecilians are vermiform, limbless burrowing amphibians without girdles in which the ancestral fish scales are retained. They have no fossil representatives. They are specialized for terrestrial life in having characteristic features, such as protrusible copulary organs. Because of burrowing habits and light avoiding tendency, they are seldom seen. They burrow in surface soil in deep places near streams and pounds. Sometimes they may swim in the streams. They feed on worms and termites. They are abundant in America, South Africa, Malaya, India and Sri Lanka.

Nieden (1955) recognized 19 genera and 55 species of Caecilians belonging to single family Caeciliidae. Some important and common caecilians are *Ichthyophis glutinosa, Hypogeophis, Typhlonectes, Geotrypectes, Dermophis, Siphonops, Rhinatrema* and *Uraeotyphlus* etc.

External Features

(1) The body is like large earthworms i.e. vermiform, elongated and eel–shaped. The head is small and depressed and indistinct. The animals are devoid of limbs and girdles.

(2) Several transverse grooves are present on the body.

(3) The tail may be very short or it is completely absent.

(4) The eyes are lidless and functionless, usually indistinct and frequently hidden under the bones of the skull, covered and concealed with the skin.

(5) There is a pair of protrusible tentacular sensory organ lodged in a special pit, which is cleaned by the secretion of the orbital gland.

(6) There is no tympanum.

(7) Vent is nearly terminal i.e. at the posterior extremity of the body.

(8) The skin is provided with numerous transverse grooves or wrinkles. The scales are present in transverse rows. The scales are however absent in some Indian genera (e.g. *Gegnophis, Typhlonectes*). Epidermal nucous glands are present in the skin whose secretion keeps the skin slimy. The squirt glands secrete irritating fluid.

Alimentary Canal: The alimentary canal can be conveniently divided into buccal cavity → Pharynx → Oesophagus → Stomach → Intestine and → Rectum. The pancreas is situated between the U–shaped loop formed by Duodenum and stomach. The liver is long and asymmetrical. It is lobed.

Respiration: Cutaneous, pulmonary and branchial respiration take place. There are two lungs. The left lung is very short but the right one is elongated. Alveoli are well developed through out. Branchial respiration takes place by gills in the larval forms.

Circulation: There are two auricles, the right auricle being larger than the left. They are separated by inter-auricular septum which is rather incomplete. Sinus venosus and conus arteriosus are also present, there is no special longitudinal valve in conus arteriosus. There are two pairs of aortic arches i.e. systemic and pulmonary arches which are joined with a small *ductus botalli*. The first branchial arch of the larval form disappears in the adult.

Nervous system: The brain is slightly modified in Gymnophiones. Owing to flexure, the hypophysis is brought beneath the medulla oblongata. Paraphysis and pinealis are present. The optic lobes are less developed. The cerebellum is very small.

Sense Organs: The tentacular apparatus situated between the eyes and the nare are important sense organs useful for exploring in the dark burrow. The Jacobson's organ or the Vomero–nasal organ is used for testing the food materials and water.

Urinogenital System

The kidney is long and narrow, extending throughout the length of the body cavity. The larval kidney is pronephros with segmentally arranged lobules having nephrostome and glomerulus. Most of them contribute in the formation of adult kidney. In caecilians, mesonephroi are more lobulated. Urinary bladder has no connection with the mesonephric duct. The bladder is a long structure.

Gonads are long in caecilians. The testes are present in the form of two strings of beads. Each bead is characterised by the presence of a number of seminiferous tubules, the strings of each side being connected by the testicular canal. The kidney duct serves the purpose of vas deferens and the ureter and open posteriorly into the side of a the cloaca. The sperm is introduced into the cloaca of the female by the reversible cloacal wall of the male, which acts as intermittent organ. The mullerian duct forms the oviduct. In the viviparous species it dilates to form the uterus. Posteriorly it opens into the cloaca.

Skeleton

The important skeletal features have been mentioned below :

1. Caecilians have a very large number of vertebrae (200-300) which are amphicoelous.
2. There is no sacrum and the 1st vertebra has no odontoid process.

3. Ribs, if present, do not fuse to form sternum.
4. Among the caecilians, the cartilage skull is known only in *Ichthyophis*. Parachordals are reduced.
5. Membrane bones form a complete roof of the skull.
6. Otics and exoccipitals are fused like quadrate and pterygoid.
7. Supra temporal and supra occipitals are absent.
8. Post frontal is present in Ichthyophis.
9. Ectopterygoid is present in Hypogeophis.
10. Parietal and frontal bones are separate.
11. Sometimes small turbinal bones may be present.
12. There is enormous development of quadratojugal arch to which squamosals fuse completely.
13. The maxilla is large and broad and it is perforated by tentacular groove.
14. Vomer is quite distinct and premaxilla is small.
15. The periotic bone is represented by Pro–otics and epiotics which fuse with the ex–occipital and parasphenoid.
16. A large prefrontal and a larger post–frontal occur.
17. Palatine and maxilla are fused in the mouth.
18. The hyoid and branchial apparatus are more primitive, and the 3rd branchial arch remains as vestige.
19. Parasphenoid and basioccipital form a large *os basale*.
20. In lower jaw, only dentale and angulare are present.

Development

Among caecilians, both oviparous (e.g. *Hypogeophis, Ichthyophis*) and viviparous forms (e.g. *Typhlonectes, Dermophis*) occur. The fertilization is internal. In *hypogeophis*, there is no larval life. Development in two genera i.e. *Ichthyophis* and *Hypogeophis* has been thoroughly worked out. The eggs are large yolky covered with albumen having twisted chalazae. The cleavage is meroblastic and the gastrulation resembles with that of reptiles. Ichthyophis breeds in the spring season. The embryo has cutaneous sense organs on the head and in the lateral side of the body. The tail fin and gills are present which degenerate later on. The hatched larva has no gills because they are atrophied. Lungs are developed and the larva rises up to take oxygen occasionally. The larva metamorphoses into the adult form and starts leading a terrestrial burrouring life. In case of hypogeophis the breeding takes place throughout the year. The lateral line sense organs and tail fin are absent. The gilled larval period expires before hatching.

In the viviparous forms (e.g. *Typhlonectes*), the development is direct in the uterus and the young ones are given birth.

Systematic Position

The systematic position of Gymnophiona is still a controversial issue. Though it has been placed in class Amphibia, but, it shows some resemblances with fishes, reptiles and also birds. Urodela (caudata), particularly *amphiuma* has been supposed to be the nearest allies of Gymnophiones by Cope. *Amphiuma* has been considered as the neotenic form of caecilians. This consideration was based on the following facts :

1. The eyes are lidless.
2. The body is eel–shaped (long and cylindrical).
3. Amphicoelous vertebrae are there.
4. Existence of tentacular groove in both the cases.
5. There is no sternum.
6. There is no spiral valve in conus arteriosus.
7. Sinus venosus opens into right auricle and the pulmonary vein opens into left auricle.
8. Foramen of Panizzae is present.
9. Ductus botalli is present in both the cases.
10. Eggs are clustered by means of threads and the female coils around them.
11. Some species of Gymnophiona resemble proteus in the absence of eyes and eye–lids.

Anuran Affinities

1. Aquatic larval forms exist in several members of Gymnophiona.
2. Larvae have tail fins which degenerate later on.

Stegocephalian Affinities

1. Retention of cutaneous calcareous scales
2. Presence of Laryngeal and Post–orbital bones
3. Acrodont teeth on jaw bones, vomers and palatine
4. Most of Stegocephalians and Gymnophiones have/had carnivorous feeding habit
5. The cranium completely roofed by bones
6. Epiotic bone present in both the groups
7. Branchial arches were primitive in character

Reptilian Affinities

Caecilians resemble some reptiles particularly snakes, inhabiting some common characters as mentioned below :

1. Elongated body without limbs
2. The head is small
3. Sternum is absent
4. Terrestrial/aquatic/burrowing forms
5. Asymmetrical lung, the right one being larger in size
6. The liver is long and asymmetrical
7. Exoskeleton of dermal scales
8. Stapes articulates with the quadrate.

Avian Affinities

1. The eggs have sufficient yolk like birds and the cleavage is meroblastic.
2. The orbit is surrounded by bones in avian fashion.

Piscine Affinities

The important piscine characters are —

1. Presence of scales
2. Pronephric kidney in the larva
3. Amphicoelous vertebra
4. Embryo coiled over the surface of yolk.

Conclusion

The avian affinities of caecilians are not very significant moreover the birds are warm–blooded 'vertebrates' much more evolved than the amphibians in many respects. Reptilian affinities are merely a matter of convergence. Several characters of Caecilians including the structure of heart and brain clearly indicate that the inclusion of this group (Gymnophiona) in Amphibia is justified. Any variation in characters is related to their burrowing habit. Now, there is general agreement that Caecilians have evolved from a common *Stegocephalian stock* and they form a group equivalent to Urodela (caudata) and Anura. They have been rightly placed in order–Apoda (Gymnophiona) of class Amphibia.

Neoteny and Paedogenesis

The term *Paedogenesis* was first used by Von Baer in 1886 to mean precocious reproduction by a larva. This precocity implies accelerated reproduction by a larva. This precocity implies accelerated development of the reproductive organs relative to the whole body. The term *neoteny* was introduced by Kollman in 1882 for retardation of bodily development compared to that of the reproductive organs as in some amphibians. The term "*foetalization*" was used by Bolk in 1926 to express the persistence of certain foetal characters or immature characters of an ancestor into the adult stages of a descendent. De Beer (1958) favours the use of Garstang's word *Paedomorphosis* which compasses the modes by which larval or immature features of ancestors become adult characters of descendants. De Beer employed the term *heterochromy* to illustrate any evolutionary change in the relative rates of development of characters during ontogeny.

The words neoteny and Paedogenesis are considered as synonymous. But, there is some difference between these two terms. The term neoteny means prolongation of the larval stage, but the term Paedogenesis refers to the stage when the larvae mature sexually and they are capable of breeding and producing young ones.

The tadpole larva of *Ambystoma* (*Amblystoma*) is of very large size. It does not metamorphose into the adult form. It grows further and attains sexual maturity. The larval form which is sexually mature but it is still a larva. It is known as *Axolotl* larva. It has been observed that when thyroid extract is administered to the larva, the metamorphosis takes place. Thus artificial metamorphosis can be easily induced in the larval forms. It has been reported that in the *axolotl* larva, thyroxine is present in the gland but it is not properly discharged into the blood stream. It may also be because of some sort of deficiency in the release of the concerned tropic hormone (TSH) from the pituitary gland. It is also possible that the desired amount of thyroxine or thyroid stimulating hormone is not secreted in the body.

That the *axolotl* larva was the larva of a Salamander was confirmed by Cuvier in 1885. It created much confusion before that.

Environmental Factors and Neoteny

Environmental factors have much to do with the metamorphosis of the larvae. There are evidences that cold climate, excess food or insufficient iodine amount may cause failure of metamorphosis. It has been pointed out that drying up of swamps, food scarcity and slightly higher temperature of the surrounding water may induce metamorphosis in the axolotl larva. The case of metamorphosis of *axolotl larva* in the United States and Mexico is quite different. In Mexico the axolotl larva permanently remains in lakes, possesses gills and shows sexual activities. But in the United States the axolotl larva of *Ambystoma tigrinum* metamorphoses into the adult form under ordinary circumstances.

Types of Neoteny

There are two types of Neoteny i.e. Artificial and Permanent. As we know, that the main hormone which induces metamorphosis is *thyroxine* of the thyroid gland. If the production of this hormone is inhibited the process of the metamorphosis can be checked and naturally the neotenic form will continue its existence. The artificial neoteny is caused by the removal of thyroid gland microsurgically or by the administration of antithyroid substances e.g. thiouracil. Permanent neoteny can be seen in the case of *Proteus anguineus* of European caves. There is no effect of any chemical, including hormones and antithyroid substances and pituitary extract.

Concluding Remarks

The concept of Neoteny or Paedogenesis accounts for the origin of certain progressive groups of animals. Weisman (1875)

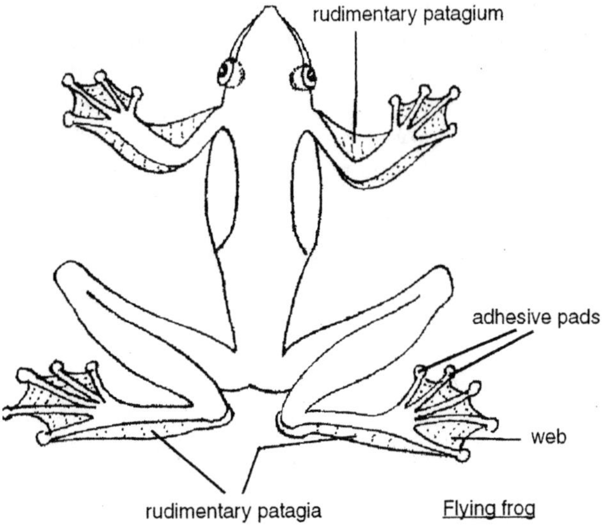

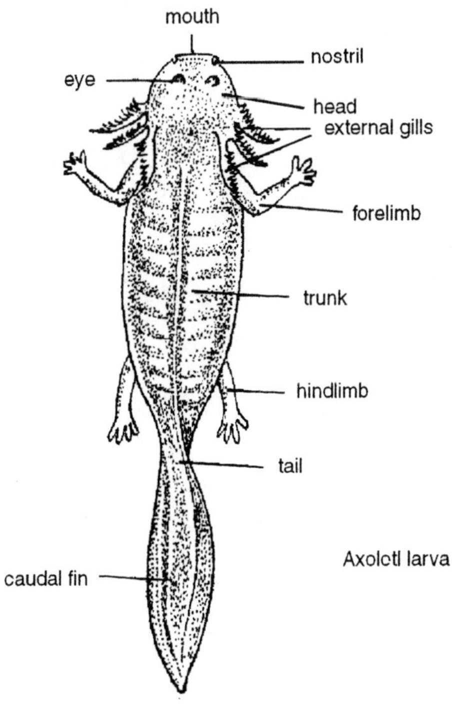

Fig.18 : Flying frog, *Rhacophorus* and *axolotl larva*

considered neoteny to be a case of retarded evolution or *atavism* where reversion of ancestral characters takes place. De Beer also realised that neoteny plays significant role in the genesis of own species. Neoteny has been considered as the fruitful source of evolutionary novelties. It is a physiological adaptation for evolutionary advantage.

Chapter 5
Class – Reptilia

Reptiles (Gr. Rept = Creeping) are cold blooded (ectothermal) terrestrial or aquatic tetrapods. They possess epidermal horny scales, some reptiles have bony dermal plates also. The skull has only one occipital condyle. In the living forms, two vertebrae form a sacrum. Limbs are pentadactyle ending in claws. Lungs are the sole respiratory organs. The ventricle is partly divided into chambers though, division is not complete. Sinus venosus is present. The kidney is that of Metanephric type. There are twelve pairs of cranial nerves. Fertilization is always internal. Extra embryonic membranes are well developed and hence they are known as amniotes. Twelve pairs of cranial nerves are present. The endoskeleton is completely ossified. Temporal vacuities or fossae may be present in the skull. Mesozoic era is known as the golden age of reptiles. *Cotylosaurians* are considered as the stem reptiles.

Rhynchocephalia

(Salient features and affinities of *Sphenodon*)

Rhynchocephalia, an Order of Class – Reptilia is considered as an aberrant group of reptiles which appeared first in Triassic and passed smoothly through the Mesozoic. It is a true diapsid reptile now represented today by one very conservative living species *Sphenodon punctatus* also called *Hatteria punctata*. It is called by the natives of New Zealand as 'tuatara' (tua = back, tara = spine). It is a lizard like reptile which grows to about 2 feet long and has well developed pentadactyle limbs adapted for walking. Gadow refers it as "the last living witness of the by-gone ages". This primitive reptile is called the "living fossil". It is a very lazy creature. It lives in the holes on the slopes of the sand hills of the shore with the petrels on amicable returns. This animal is rigidly protected by law. It emits a soft frog like croak. It feeds on insects, crustaceans and worms.

Distribution

Sphenodon was formerly known on the main islands of New Zealand, but is now restricted to some small islands in the Bay of Plenty and appears to be on the verge of extinction.

General anatomical features

(1) The lizard like body of *Sphenodon* is about 2 feet long and the tail measures about 1/3 of the whole length of the body. The animal is nocturnal.

(2) The body and tail are laterally compressed and a series of spines are dorsally present (along the middle line).

(3) Limbs are pentadactyle, legs are short and primitive in nature. Sometimes 10-11 carpels are present.

(4) Hind limbs are typically plantigrade.

(5) The upper surface of the body is covered with small granular scales and the lower surface is covered with transverse rows of large squarish scales.

(6) The head is large and there is prominent ridge over the eye.

(7) There is a foramen above the outer and one above the inner condyle of the humerus.

(8) There are eleven carpal elements in all, four in the proximal row, two centrals and five in the distal row.

(9) In the tarsus the *tibial* and *fibular* elements are quite distinct, though firmly united. The intermedium and the centrale are firmly fixed to the tibiale. There are three tarsal bones.

(10) Teeth are pleurodont and homodont like reptiles but they are monophyodont. The teeth are pointed, triangular, laterally compressed and arranged in two parallel rows, one along the maxilla, and the other along the palatine. The teeth of the lower jaw bite in between the two upper rows, all the rows becoming worn down in such a way as to form continuous ridges. Each premaxilla bears a prominent, chisel–shaped incisor, represented in the

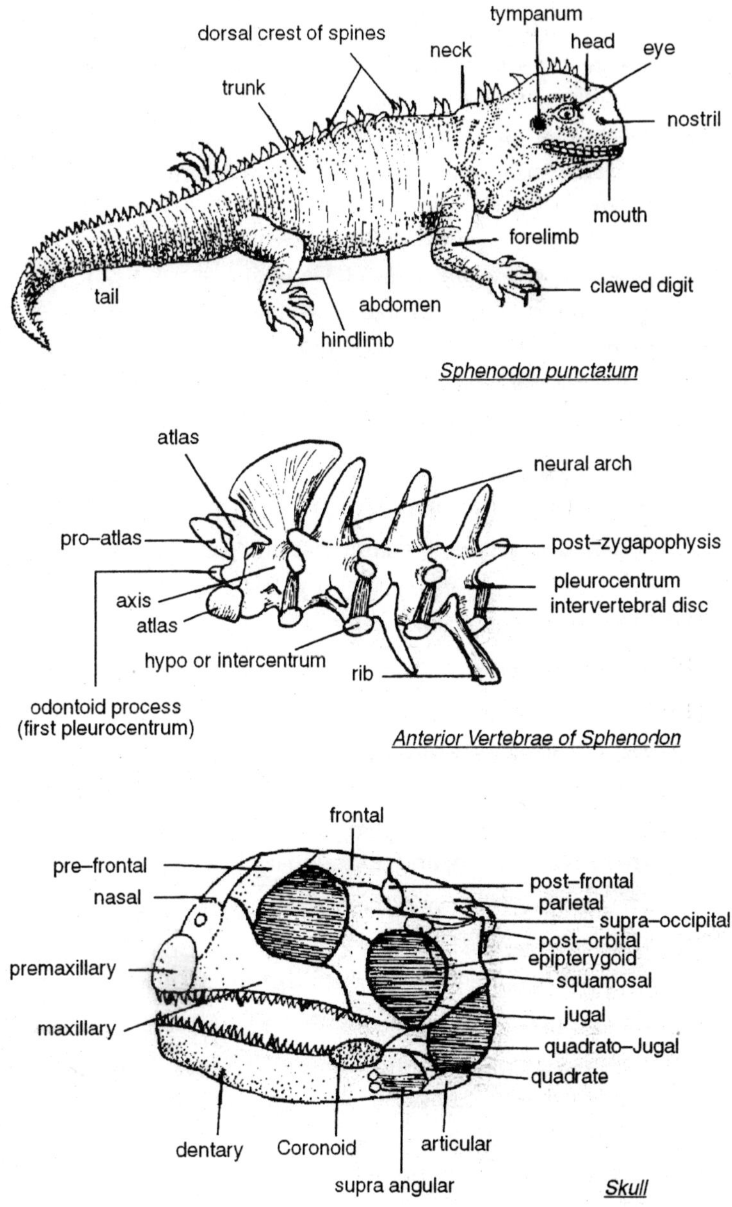

Fig.19 : External features, anterior vertebrae and skull of young one

young animal by two pointed teeth. In the young *Sphenodon* a tooth has been found on each vomer — a condition unusual among reptiles.

(11) The pubes are united in a Symphysis, infront of which is a cartilaginous epipubis. A large oval foramen intervenes between the ischium, and the pubis. A cartilaginous hypoischium is attached to the ischia behind.

(12) In the skull, there is a complete lower temporal arch. The quadrate is immovably fixed, wedged in by the quadrate–jugal, squamosal, and pterygoid. The premaxillae are not fused together but separated by a suture. There is a broad palate formed by the plate–like vomer, palatine and pterygoids. The pterygoids meet in the middle line and extend forwards to the vomers. There is an epipterygoid extending from the parietal to the pterygoid and quadrate.

(13) Sphenodon retains the large parietal foramen, in which, is the non-functional median pineal eye.

(14) There is no antorbital foramen.

(15) The anal opening of Sphenodon is transverse.

(16) There is no copulatory organ (Penis) in the male.

(17) The vertebrae are amphicoelous and fully ossified and intercentra are present.

(18) The ribs are single headed and have *uncinate processes*. There is a sternum and abdominal ribs.

(19) Chevron bones occur in the candal region of the vertebrae.

(20) There is a median bone, the *Pro–atlas*, intercalated between the atlas and the occipital region of the skull.

(21) Caudal ribs become fused with the vertebrae.

(22) The lateral temporal fossa is bounded below, by an inferior temporal arch composed of jugal and quadrato-jugal (Palato-quadrate).

(23) *Jacobson's organ* is present in a rather primitive form.

(24) A urinary bladder is present.

(25) There is a T-shaped interclavicle.

(26) Coracoid is without fenestra.

(27) It lays 10-13, leathery shelled eggs which are buried few inches below the surface of the soil. The young ones are hatched 13 months later.

Causes of its long survival

There are three important causes of its long survival. These are —

a) It was free from its enemies because the placentals and other large terrestrial animals were introduced in New Zealand quite lately.

b) Slow rate of metabolism and conservation of sufficient energy.

c) Lethargic habit and long period of incubation of the fertilized eggs.

Why called a living fossil

The living fossil is an animal that has survived beyond its era. It is the one species or such a group of species that has continued existence when the other closely related contemporary species have become extinct. *Sphenodon* is the sole survival of the order *Rhynchocephalia*. Its anatomical features indicate several primitive characters such as —

1) The skull bones are shaped and disposed in the manner of extinct group of reptiles.

2) Teeth are fused to the jaw bones.

3) Presence of pineal foramen.

4) There are several characters (discussed in affinities) which resemble that of Dinosaurs and other extinct reptiles.

5) It resembles with the ancient *Homaeosaurs* in several features except the presence of uncinate process of the ribes in Sphenodon.

Thus, the term *'living fossil'* is justified for *Sphenodon*. This animal which has been called the "Voice of the Past" has been given absolute protection by law in New Zealand.

Affinities

Sphenodon (Order – *Rhynchocephalia*) possesses several characters which are present in different groups of animals. So, actual position of *Sphenodon* in the animal kingdom was somewhat controversial for some time. Its affinities with different groups have been discussed below :

Affinities with amphibia

The circulatory system of Sphenodon has some resemblances with Caudata or Urodela. These resemblances are —

1) Aortic arches arise from a short common stalk which is comparable to *conus arteriosus* of the amphibians.
2) Presence of *ductus arteriosus* and *ductus caroticus*
3) The pattern of distribution of blood vessels almost similar

But, these resemblances are mainly because of the reason that *Sphenodon* is an animal of primitive nature. Because of the presence of several reptilian features which have been discussed below, amphibians affinities can not be accepted.

Affinities with Dinosaurs

There are some resemblances with the dinosaurs. These are—

1) Diapsid condition of the skull i.e. the skull has two vacuities (fossae).
2) Fixed quadrate bone
3) Abdominal ribs
4) Uncinate process of the ribs.

But, there are some differences also — e.g.

1) In Dinosaurs, double headed ribs are present – not single headed like *Sphenodon*.
2) Dentition is thecodont in dinosaurs
3) Absence of clavicle and interclavicle
4) Absence of Parietal foramen
5) Absence of third eye elements

So, *Sphenodon* can not be linked with the dinosaurs, however, its reptilian ancestry is evident.

Chelonian affinities

The following are the important resemblances with the chelonians :

1) Fixed quadrate bone
2) Ribs of caudal region fused with the vertebrae
3) Pecten not present in the eye
4) Urinary bladder present

The differences with chelonians include the following —

1) *Sphenodon* is terrestrial but chelonians are aquatic.
2) Teeth absent in the chelonians, only horny beaks present
3) Parietal foramen in chelonians
4) Unpaired vomer
5) No sternum in chelonia
6) Cloacal opening longitudinal in chelonia but transverse in Sphenodon
7) Males have penis in chelonia
8) Opening of oviduct on ventral side

Affinities with Crocodilia

The following are the important resemblances with Crocodilia—

1) Diapsid condition of the skull
2) Fixed quadrate
3) Presence of Pro–atlas
4) Presence of abdominal ribs
5) Fusion of caudal ribs with vertebrae
6) Uncinate process of ribs
7) Presence of chevron bones
8) Cochlear process tubular

The differences between the two are :

1) Thecodont dention in crocodiles
2) Single nostril in Crocodile but double in *Sphenodon*
3) Procoelons vertebrae in Crocodile and amphicoelous in *Sphenodon*
4) Clavides not present in Crocodilia

5) Pecten present in Crocodiles
6) Penis present in male Crocodile

Affinities with Lacertilia

The important resemblances are the following —

1) The body plan is similar
2) Pro–atlas present
3) Amphicoelous vertebrae in certain geckos
4) Single headed ribs
5) Chevron bone present in both
6) Structure of the respiratory organs
7) Parietal organ common
8) Cloacal glands present

However, there are some differences also, which are —

1) Quadrate fixed in Sphenodon
2) Most lizards have procoelous vertebrae
3) Rami of the jaw united by symphysis in lizards
4) Erect ilium in sphenodon
5) Clavicles and interclavicles present in Sphenodon
6) Absence of conus arteriosus in Lacertilia
7) Presence of Pecten
8) Uncinate process of the ribs absent in Lacertilia
9) Copulatory organs present in Lacertilia
10) Presence of lower temporal arch

Conclusion

Thus, we see that Rhynchocephalia retains many primitive characters and resembles chelonians, crocodilia and dinosaurs in many features. But it is convincing that *Sphenodon* is more closely allied to the lacertilia than any other group discussed above. Because of the occurrence of certain peculiar features it seems justified that the Rhynchocephalia should be placed as a separate order of class–Reptilia as suggested by Romer.

Chelonia

Chelonia is an important order of class – Reptilia. It comprises tortoises, terrapins and turtles. Fresh water tortoises or turtles are included in the family *Emylidae* and the marine turtles belong to the family *Chelonidae*. In Britain, the land living Chelonians are called tortoises and the marine Chelonians are often called turtles. In the United States, the term turtle is used as a general name for all the chelonians but sometimes the terrestrial forms are called tortoises. The name *terrapin* was used for a variety of edible salt water and brackish water chelonians of North America. The chelonians are called "reptiles in box". Tortoises are essentially vegetarian. Among the land tortoises, the genus *Testudo* is very common and they are widely distributed in the tropical and temperate regions of Asia, Africa and Europe. They are not present in Australia and Polynesia. *Testudo elegans* is found in India and Ceylon. During the summer season it goes into water but during the cold seasons it hibernates. The genus *Trionyx* is commonly known as *tortoise* or *soft river terrapin*. They are distributed in the rivers of Asia, Africa and North America. *Trionyx gangeticus* is found in the rivers of Northern India. The Green Turtle known as *Chelone mydas* is a marine turtle which is in danger of extinction. It is found in Indian, Pacific and Atlantic Oceans. The European tortoise, *Emys orbicularis* is strictly carnivorous. The body in chelonians is entirely covered with firm shell the upper part of which s known as the *Carapace* and the under part is called *Plastron*. Both carapace and plastron are formed by the welding together of polygonal dermal plates. Chelonians are also peculiar in having horny jaws resembling those of birds. Tortoises survive to a great age. *Testudo gigantea* kept in captivity on Mauritius lived to a known age of 152 years. Chelonians have many peculiar and rare characters which have been discussed later on.

External Features

The body is divisible into head, neck, trunk and tail. The trunk is enclosed in a bony shell, composed of dorsal *Carapace* and ventral *plastron* – which may or may not be covered with shields. The head, neck limbs and tail can be withdrawn into the shell and in that condition it looks like a box. The head is dorso–ventrally flattened. The mouth is large and the teeth are absent. The horny jaws are present like beaks of a bird. At the extremity of the snout the external nares are placed together. One eye is present on either side of the head. It is guarded by *upper eye lid, lower eye lid* and the *nictitating membrane.* There are two pairs of limbs which are pentadactyle. The skin is thin and smooth on the head but it is thick, tough, scaly and wrinkled over the rest part of the body which are exposed.

Shell

As pointed out above, the dorsal part of the shell is known as *Carapace* and the ventral part is known as the *Plastron.* The carapace and plastron are bound together on each side by bony bridges or ligaments. The carapace is a convex structure consisting of thick bony plates. It consists of a series of *median plates* bounded laterally by *costal plates.* The *costal plates* are peripherally surrounded by *marginal plates.* The median plates are also known as neural plates and they are fused below with the vertebrae. The costal plates are also fused to the underlying ribs. The median series consists of a large *nuchal* plate (anterior and dorsal in position) and eight *neural plates.* There are eleven marginal plates on each side and there are eight pairs of costal plates.

The plastron consists of nine dermal plates :

These are — One pair of *epiplastron* (anterior most),
One pair of *hyoplastron,*
One pair of *hypoplastron,*
One pair of *Xiphiplastron* (Posterior most)

and a median *entoplastron* surrounded by epiplastron above and hyoplastron below, thus being situated at the junction between all four pieces. The epiplastron is considered as the clavicle and the entoplastron represents the episternum of other tetrapods. The hyoplastron, hypoplastron and xiphiplastron are regarded as homologous with gastralia (abdominal ribs) of other reptiles.

Shields or Scutes

Carapace and Plastron are covered by several symmetrically arranged horny shields which do not correspond in number or arrangement to underlying bony plates. Altogether thirty seven shields cover the carapace — they are five vertebral, four pairs of costal and twelve pairs of marginal shields. The nuchal shields lie in front of first vertebra. They are very small and sometimes may be absent. Posteriorly the last pair of marginal shields may be united to form single *pygal shield.* Similarly, the plastron is

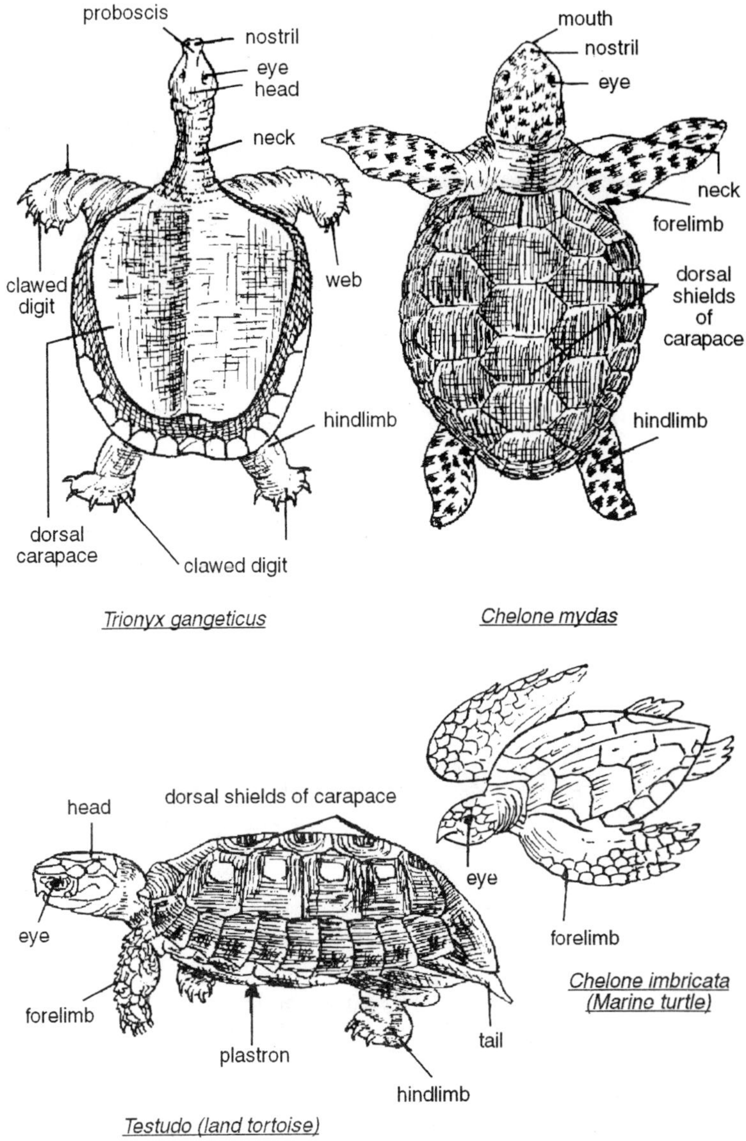

Fig.20 : Some members of Chelonia

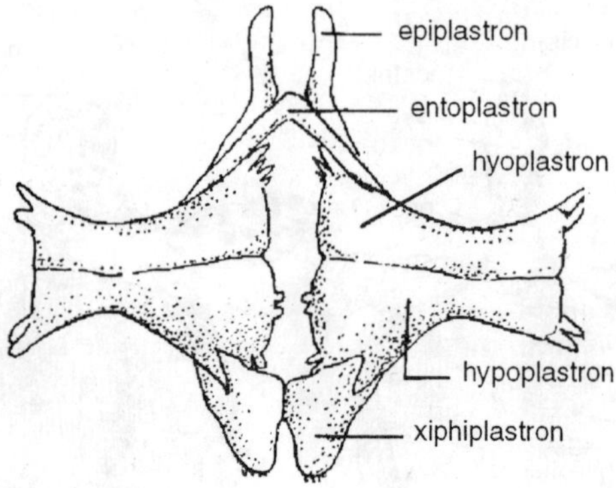

Fig. – Plastron of *Trionyx*

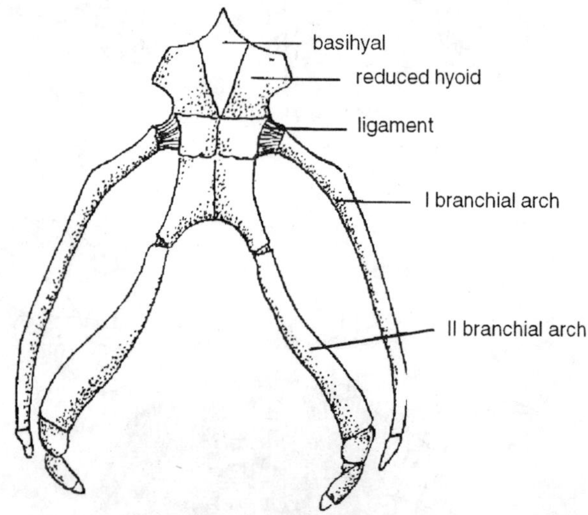

Fig. – Hyoid apparatus of *Trionyx*

Fig.21 : Plastron and Hyoid apparatus of *Trionyx*

covered with twelve pairs of shields — i.e. *paired gular, paired humeral, paired pectoral, paired abdominal, paired femoral* and lastly *paired anal shields*. Sometimes, and *intergular* may be present.

Skull

The bones of the skull are immovably united to form complete uninterrupted roof of the membrane bones without temporal vacuities (fossae), and hence it is of *anapsid* type. The foramen magnum is bounded by *basioccipital,* two *exoccipital* and *supra–occipital.* In certain cases the basioccipital is excluded from the foramen magnum because, this bone and the occipitals participate in the tripartite occipital condyle. Membranous bones of frontal segment consists of *prefrontals, frontals* and *post frontals.* Prefrontals join ventrally with *vomer* which is a single vertical bone and the *palatine.* The palatines are firmly united to *Pterygoids* and *basisphenoids.* The *post–frontals* extend posteriorly and ventrally to join *squamosals, quadratojugal* and *jugals.* There are five bones i.e. *maxilla, prefrontal, frontal, post orbital* and *jugal* which surround the orbits. In the otic capsule the *epiotic* and *opisthotic* are ossified. The opening between *pro–otic* and *opisthotic* is called the *auditory meatus.*

The jaws are without teeth and are covered by horny sheath. In the skull, there is only a lower temporal arcade formed by jugal and quadratojugal. *Pre–sphenoid, orbitosphenoid* and *alisphenoid* are not perfectly ossified. The cartilage bone of the upper jaw, *quadrate* dorsally joins squamosal and anteriorly it joins .quadratojugal. Vomers and Palatines contribute in the formation of external and internal nares. The mandible is made up or *articular, angular, supra–angular, coronoid, dentary* and *splenial bones.*

Vertebrae

In turtle there are 36–55 vertebrae in all. The number of vertebrae has been indicated below :

Cervical	8
Thoracic	10
Sacral	2
Caudal	16–35

The first cervical vertebra is known as *atlas.* It is composed of one ventral and two darso–ventral pieces. The second vertebra is known as *axis.* At the anterior end of the centrum it bears a projection known as *Odontoid* process. The thoracic vertebrae bear ribs and are firmly associated with the carapace by ribs, transverse processes and neural arches. The sacral vertebrae also fuse with the carapace. The caudal vertebrae are freely movable. They bear rudimentary ribs and they are amphicoelous.

Pelvic girdle

The pelvic girdle is made up of *ilium, ischium* and *pubis.* A depression at the junction of these three bones to which the head of the femur is attached, is known as *acetabulum.* The dorsal ilium is directed backwards and they are attached to the carapace. The posterior ischium and anterior pubis with *prepubic* process and epipubic cartilage are also quite distinct. The pubes and ischia which lie side by side in the middle line are fused and form the two symphyses. *Obturator* foramen is also present to enclose blood vessels and nerves. The entire pelvic girdle is arched by the carapace.

Pectoral girdle

In the ventral side, a projection is formed from the *Scapula* which is known as *proscapular process.* In the posteroventral side, the *Coracoid* bone is present. The *Scapula* is present in the dorsal side. At the junction of all these bones the *glenoid cavity* is present. In this cavity, the head of the humerus articulates. Two membrane bones, paired clavicles (epiplastra) and inter–clavicle (entoplastron) have fused with the plastron.

Hyoid apparatus

The hyoid apparatus supports the tongue. The hyoid apparatus consists of a cartilaginous *basihyal* and the two pairs of carnua which do no reach the *cranium.* The pair of anterior cornua represents the first branchial arches and the posterior pair represents the second branchial arches. In the postero–lateral part of the basihyal the reduced *hyoid* is present.

Ribs

The ribs are confined to the dorsal side of the body and are fused to the costal plates to form the carapace. The cervical ribs are absent. The thoracic vertebrae possess the ribs. The rib has a single head, *Capitulum* for articulation with vertebra. The ribs of sacral vertebrae unite with the pelvic girdle. Anterior caudal vertebrae have ribs, like very small projection. The sternum is absent in chelonia.

Limb–bones

In the fore–limb the important bones are the *humerus, radius* and *ulna* and bones of *wrist, palm* and *fingers.* In the humerus the proximal part is the head. The distal end of the humerus possesses the *olecranon process* and *olecranon fossa.* The distal part of the humerus articulates with the radius and ulna of the fore arm. The distal parts of the radius and ulna articulate with carpals. In the wrist, at the distal end of ulna, *ulnare* and *intermedium,* bones are present. The central position is occupied by the fused median bones *radiale* and *centrale.* In the distal row there are five *carpals* which are attached with the metacarpals. The palm comprises five *metacarpals.* There are five fingers (Pentadactyle limb). The first possesses two *Phalanges* and in the rest, there are three phalanges each. The last phalanges terminate into claws.

In the *hind limb* the first bone attached with pelvic girdle with its *head* portion is the *femur.* The distal portion of this bone bears condyles and articulates with *tibia* and *fibula* of the shank. The femur is the bone of the thigh. In the foot portion, the *ankle* is made up of *tibiale, intermedium, centrale* and *fibulare* bones, fused together. There are five *metatarsals* in the *sole* portion. The toes comprise the bones of the phalanges. The first toe has only two bones but the rest four toes have three phalanges each. The phalanges terminate into horny claws. It may be pointed out that in the reptilian limb there is the occurrence of an intratarsal joint.

Digestive System

Chelonians are both *Carnivorous* and *herbivorous.* The digestive system can be discussed under the two heads – the *alimentary canal* and the *digestive organs.*

Alimentary Canal

The mouth leads into the mouth cavity which is devoid of teeth. A tongue is present at the floor of the mouth cavity. The *glottis* is a longitudinal slit at the base of the tongue. At the angle of the jaws the openings of the *Eustachian tube* are present. The *pharynx* leads into the *oesophagus* through the *gullet.* The oesophagus communicates with the sac like stomach. The *gastro–hepatic ligament* attaches the stomach with the liver lobe. The stomach leads into the *duodenum* and the latter is continuous with the ileum. The *hepato–duodenal ligament* makes attachment of the duodenum with the liver lobe. The *pancreas* is present in this ligament. The ileum is coiled and suspended with the help of *dorsal mesentery.* The part of the mesentery which supports the stomach is known as *mesogaster* and the part supporting the duodenum is called the *mesoduodenum.* The ileum leads into the large intestine, also known as *colon* which finally opens into the rectum. At the junction of large intestine and the small intestine, the *ileocaecal valve* is present. Now, the rectum opens in to the *cloaca* and the latter opens outside through the *cloacal aperture.*

Digestive Glands

The liver, the pancreas, gastric glands and intestinal glands are the main digestive glands.

The left lobe of the liver is attached to the stomach and the right lobe is attached to the duodenum. The *gall–bladder* is attached on the surface of the right liver lobe. The bile duct comes from the liver and enters into the duodenum. Bile is secreted by the liver and makes the medium alkaline. The pancreas is present in the hepato–duodenal ligament. The pancreatic duct opens into the duodenum. The pancreatic juice contains enzymes which are of great use in digestion. The gastric glands of the stomach secretes gastric juice and the intestinal glands secretes intestinal juice. All are important for the digestion of food materials.

Respiratory System

At the anterior end of the snout there is a pair of *external nares.* These apertures lead into the *nasal passages* which open into the internal nares. At the base of the tongue there is a slit like aperture known as the *glottis.* This glottis leads into the *larynx.* The larynx communicates with the *trachea* or *wind pipe,* the walls of which have *cartilaginous rings.* The rings are not complete in the posterior region of the trachea. In some turtles, the trachea may be convoluted which is merely an adaptation for quick reduction of the size of the neck to preserve it in the shell, when necessary. The trachea divides into two *bronchi.* The bronchus of one side enters into the lung of its corresponding side. The lungs are the main organs of respiration. These are spongy and richly supplied with blood capillaries. The cavity of the lung is divided into small spaces or alveoli. As the ribs are fused to the compact shell, volume change in the shell is caused by the contraction of two sets of muscles. During respiration, the breathing movements transmit pressure change on the visceral organs.

Circulatory System

The Heart

The heart is a thickwalled conical organ lying in the pericardial cavity. It has three chambers – two *atria* and a *ventricle.* The entire heart is surrounded by a membrane known as *Pericardium*. The atrial septum is complete and it continues forward into the ventricle partially separating the latter into the arterial and venous halves. A large chamber known as *sinus venosus* is attached to the right auricle. It is a centre for the precaval and postal caval veins. The *conus arteriosus* is generally very much reduced.

The inter–auricular septum divides the atrium into right and left chambers. The wall of the atrium is thin because it has not to pump out the blood. Deoxygenated blood from sinus venosus enters into the right auricle through an aperture. This aperture (opening) is guarded by muscular flaps. Similarly, the pulmonary veins open into the left auricle through an aperture. Both the auricles open into the ventricle by *auriculoventricular apertures*. The ventricle walls are thick and have muscular ridges projecting into the cavity of the ventricle. The ventricle is also incompletely divided into two parts. As a result the left and the right ventricles are in contact with each other. From the ventricle three arches are given out and they are guarded by semilunar valves.

Working of the Heart

The blood is collected from different parts of the body through the veins and it is poured into *sinus venosus* through major veins i.e. precaval and post caval veins finally. From sinus venosus the blood passes into the right auricle. This impure blood goes into the right ventricle and from there the blood goes into the lungs through pulmonary arch, and this arch gives rise to the pulmonary arteries. The pulmonary veins carry pure blood to the left auricle. From the left auricle, the blood reaches the left ventricle. When the muscular walls of the ventricle contract the blood is pumped into the systemic arches. Mixing of pure and impure blood takes place to some extent.

Arterial System

The arteries carry blood to different parts of the body to supply blood to various tissues. From the ventricle three arches are given out and they are : *right systemic, left systemic* and *pulmonary arches*. These arches are guarded by semilunar valves at their bases.

The right systemic arch gives off a *branchiocephalic* (innominate) artery. It is a common trunk formed by *subclavians* and *carotids*. The heart is supplied by the *coronary arteries*, arising from the base of the brachiocephatic artery. The brachiocephalic artery divides into *right sub–clavian, left sub–clavian, right common carotid* and *left common carotid* branches. The sub–clavian artery gives off a *ventral cervical artery*. Beyond the ventral cervical artery, the sub–clavian is known as *axillary artery*. Three arteries are given rise by the axillary artery – they are *dorsal cervical, first intercostal* and *vertebral*. The axillary artery, near the origin of the first intercostal and vertebral arteries is known as brachial artery. The common carotid artery divides into *external* and *internal carotid arteries.*

The ventral cervical artery supplies blood to oesophagus, trachea, neck muscles, thymus and the thyroid glands. The dorsal cervical goes towards the neck region. The first intercostal joins the margino–costal supplying blood to the carapace. The vertebral artery goes to vertebral column. The blood is supplied to the fore limbs by the brachial artery. The external and internal carotid arteries supply blood to the head region.

From the left systemic arch three large arteries arise, they are — *Gastric artery, Coeliac artery* and the *Anterior mesenteric artery*. The coeliac artery divides into *anterior* and *posterior pancreatoduodenal* arteries. As the names indicate, the gastric artery supplies the stomach. The pancreatoduodenal arteries supply blood to the stomach, pancreas, duodenum, liver and gall bladder. The anterior mesenteric artery supplies blood to the mesentery.

The right and left systemic arches join together to form the *dorsal aorta* which runs posteriorly in the mid–dorsal line. It gives off *renal arteries* and the *genital branches*. The dorsal aorta, behind the kidney forms two arteries, the *left* and *right common iliac arteries*. From the common iliac artery, arise the *external* and *internal iliac arteries*. Now again, the external iliac artery gives rise to the *sciatic artery* and *femoral arteries* supplying blood mainly to the hind limbs. The branches of internal iliac artery reach the gut, bladder and genital organs. Behind the kidney, the *epigastric artery* also arises from the dorsal aorta and it gives two branches i.e. *anterior* and the *posterior* branches. The former joins the marginocostal artery and the latter supplies the fat bodies. The renal arteries already mentioned, supply the kidney and the genital branches going into the gonads.

The pulmonary arch gives rise to the *right* and *left pulmonary arteries* which supply the lungs. There is a connection between the right and the left fourth aortic arches through an aperture – the *foramen of Panizzae* and through this aperture some admixture of arterial and venous blood can occur.

Venous System

There is a system of veins through which blood is collected into the heart from different parts of the body. There are some major veins like *Precaval vein* and the *Post caval vein*. The right precaval vein receives — a *thyroscapular vein, internal jugular vein, sub–clavian vein* (axillary vein) and *sub–scapular vein*. The *sub–clavian vein* is formed by the union of *external jugular vein* and *brachial vein* and both of them collect blood from the fore limbs, neck region and the head. The sub–scapular vein collects blood from the scapular muscles. The thyroscapular vein collects blood from the thyroid gland and the shoulder muscles. The internal jugular vein collects blood from the oesophagus. The deoxygenated blood is collected into *sinus venosus*. The *right* and *left hepatic veins* coming out from the lobes of the liver, also join the precaval vein.

In the medial side of the kidneys, two veins unite to form the *Post caval vein*. It enters the liver and then emerges out to open into the sinus venosus finally. The two ventral *abdominal veins*, are connected with each other by a cross–bridge. They receive the *pericardial veins* and *pectoral veins* and enter into the liver. The ventral abdominal vein receives the *hypogastric vein* and the *pelvic vein*. The hypogastric vein receives blood from the urinary bladder and the pelvic vein receives blood from the pelvic region. Now, *crural vein* from thigh and *femoral vein* from legs also join it. It is now called the *iliac vein* to which the *epigastric vein, ischiatic vein* and *cloacal vein* join.

From the antero–dorsal surface of the iliac vein, the *renal portal vein* continues forward. It reaches the kidney and before reaching there it receives the *vertebral branch* and the *intercostal branch*. The intercostal branch anastomoses by means of the *marginocostal vein*.

The hepatic portal vein is a large vein near the gastro–hepatoduodenal ligament and it is formed by a group of veins. In the liver, the hepatic portal vein forms a complex pattern of sinuses and capillaries before entering the heart. A capillary circulation occurring in the course of a vein is known as a *portal system* and when it occurs in the liver, it is called *hepatic portal system*. Thus, the blood is brought to the liver by the *portal vein* — a part of the original omphalomesenteric – the capillary vessels, and the bases of both omphalomesenterics, known as *hepatic veins*. They convey blood to the heart from the liver. The hepatic portal vein receives several vein e.g. *gastric vein*, (from stomach), *cystic vein* (from gall bladder), *pancreatic veins* (from pancreas), *duodenal vein* (from duodenum), *splenic vein* (from spleen), and *mesenteric veins* (from mesentery). The hepatic vein after emerging from the liver, enters the *sinus venosus.*

Nervous System

The central nervous system comprises the brain and the spinal cord. The brain is covered by means of *meninges.* The inner membrane of meninges is known as the *piamater* which is a thin and vascular layer, and the outer membrane is called the *duramater* which is a tough layer. The space between the duramater, and the piamater is called the *sub–dural space.* A space is also present between the duramater and the cranial lining. This space is called the *epidural space.* Both the above mentioned spaces are filled with *cerebro–spinal fluid* which protects the brain.

As usual, the brain can be conveniently divided into three parts—

(i) *The fore–brain* (ii) *The mid–brain,* and

(iii) *The hind–brain*

The fore–brain is divisible into –

(a) The olfactory lobes

(b) The cerebral hemispheres and

(c) The diencephalon

Now, these are described below :

The olfactory lobes are present in the anterior most part of the brain. These are oval structures. Each lobe is attached to *olfactory bulb* at the extremity by olfactory tract. Like other higher vertebrates the cavity of the olfactory lobe is called *rhinocoel.* The cerebral hemispheres are well developed and the intercerebral fissure separates the two hemispheres. The cavities inside the cerebral hemispheres are known as *lateral ventricles.* The floor and lateral walls of the lateral ventricles are called *corpora striata* and roof covering the ventricle is known as *pallium.*

The cerebral hemispheres are posteriorly followed by a narrow area, the *diencephalon.* Its cavity is called the *diocoel* or *third ventricle.* In the dorsal side of diencephalon there are two projections known as *paraphysis* and *epiphysis* (pineal body) and in the ventral side, the *infundibulum* and (attached to it) the *Pituitary body* are present.

In the ventral part of the cliencephalon lies the *optic chiasma* which is formed by the crossing of optic nerve fibers. The roof of the diencephalon possesses the *anterior choroid plexus* which is a vascular tuft formed of non–epithelial cells and vascular piamater. The diocoel communicates with the lateral ventricles through *foramen of Monro* :

A pair of large *optic lobes* are present in the mid–brain. The cavity of the optic lobes is known as the *optocoel*. It communicates with the diocoel anteriorly and with the *fourth ventricle* (i.e. with the cavity of the cerebellum) posteriorly. The fourth ventricle is also known as *metacoel*. The narrow cavity between the optocoel and the metacoel is known as *iter*. Between the diencephalon and optic lobes, there is a band of nerve fibres known as *posterior commissure*.

The hind–brain comprises the *cerebellum* and *medulla oblongata*. *Metacoel* continues into the medulla oblongata also. In turtle, the cerebellum is a small lobe. The medulla oblongata is continuous with the spinal cord. The roof of the metacoel forms the *posterior choroid plexus*.

Spinal Cord

The Spinal cord is long and dorso–ventrally flattened tube. It is situated in the neural canal portion of the vertebral column. The *dorsal fissure* is a shallow groove in the mid–dorsal line of the Spinal cord, and similarly, in the ventral side also, the *ventral fissure* is present. There is a narrow central cavity in the spinal cord known as the *central canal*. This cavity is filled with *cerebrospinal fluid*. The central canal communicates with the *fourth ventricle* anteriorly but it closed posteriorly. There are two types of nervous tissues in the spinal cord. In the inner portion the *grey matter* is present but in the outer portion the *white matter* is there. The grey matter is H–shaped with dorsal and ventral columns or cornua from which the dorsal and ventral roots of spinal cord respectively arise. The grey matter is composed of nerve cells and non–medullated nerve–fibres. The white matter is mainly composed of medullated nerve fibres and it is divisible into *dorsal, ventral* and *lateral funiculi*. Sensory and motor impulses are conducted by the spinal cord. The spinal cord is a centre of reflex action.

The *Peripheral nervous system* includes the nerves originating both from the brain and the spinal cord. The nerves originating from the brain is known as *cranial nerves* and those originating from the spinal cord is known as the spinal nerves. There are twelve pairs of cranial nerves in turtles. The first 10 pairs are like the amphibians, but there are two additional pairs. The 11th pair is known as the *spinal accessory* and the 12th pair is *hypoglossal*.

The number of *spinal nerves* corresponds with the number of vertebrae in the vertebral column, the first pair arising from between the skull and the atlas vertebra. A few spinal nerves combine to from the *brachial plexus* and the *lumbar* (or *sacral*) *plexus* and then they enter the forelimb or hindlimb as one or more nerves of larger size. Ventral branches of the last four pairs of cervical nerves and the first dorsal nerve (nerve of trunk) unite to form the brachial plexus. A large *dorsal spinal ganglion* gives nerve branches. The spinal nerve formed by the union of dorsal and ventral roots divides into three branches and the branch possesses sensory and motor fibres. They are known as *ramus dorsalis, ramus ventralis* and *ramus visceralis*. The ramus dorsalis goes to the skin and muscles of the dorsal side, the ramus ventralis goes to sides and ventral parts of the body and the ramus visceralis goes to the roof of the coelom near the insertion of misentery. There are two groups of nervous elements in the sensory and motor nerves. These are *somatic sensory* and *somatic motor fibres* which are distributed to skin, external sense organs and voluntary muscles. The *visceral sensory* and *visceral motor* fibres supply the visceral organs. The dorsal and the ventral rami contain somatic fibres and a few visceral fibres but the visceral rami are composed of only visceral fibres.

The dorsal roots are mainly composed of sensory fibres and the ventral roots possess only motor fibres. The dorsal roots and their fibres carry nervous impulses to the spinal cord and they are called *afferent*, but the impulses in the ventral roots are carried in the opposite direction by the *efferent fibres*. The ventral roots are actually the *motor roots*. The ventral rami of the 8th, 9th and 10th dorsal nerves and two *sacral* nerves form the *lumbo–sacral plexus* particularly for the hind limbs.

In the autonomic nervous system there are two autonomic trunks, one in each side of the vertebral column. Each trunk is connected with all the spinal nerves by a number of ganglionic enlargements. Both sensory and motor functions are performed by this system and also, the involuntary actions are regulated. The autonomic nervous system have been divided into *sympathetic (thoracolumbar)* and *parasympathetic (craniosacral)* types. In the sympathetic type, the thoracic and lumber spinal nerves are included. They are responsible for accelerating heart action, activating sphincter muscles, and, inhibiting smooth muscles of the digestive tract. In the parasympathetic system oculomotor nerve, the facial nerve, glossopharyngeal nerve the vagus and few sacral nerves are physiologically active. The cephalic part of the autonomic nervous system is well developed and is connected with several ganglia, e.g. *Cervical, Sphenopalatine, Otic* and *Submaxillary*. They receive branches from the cranial nerves. Besides the visceral sensory and motor elements, there are purely sympathetic elements also. The fibres of ramus visceralis arise either from ganglion cells in the dorsal ganglion or from the lateral column (cornu) of the spinal cord. They develop axons and dendrites which interlace with motor and sensory ganglion cells of the chain ganglia.

Sense Organs

The sensory organs are known as *receptors*. They receive informations from various parts of the body. Normally the sensory cells arise from the ectoderm, there being very few exceptions.

Olfactory–receptors

The sense of smell and taste are very closely associated and sometimes they are spoken as chemical senses. Of course there are histological differences between the tissues of these concerned organs. The *nasal chambers* are responsible for sense of smell. The projections in the nasal cavity are known as *conchae*. They are developed in the lateral wall of the cavity, however these are weakly developed in turtles. The nasal cavity opens into the mouth cavity through the internal nares. In the nasal passage, there is a sensory epithelium. The Jacobson's organ is absent in chelonia.

Gustato–receptors

The taste buds are the organs responsible for carrying the sense of taste. These organs are restricted to the cavity of the pharynx. Its presence on the tongue would not have been convenient owing to keratinous nature of the tongue. The structure of the taste buds is the same as in other tetrapods.

Photo–receptors

There is one pair of eyes and they are the real photo–receptor organ. They are very small in size and they are similar to fish in structure. The cornea is somewhat flattened. Spherical lens and retinal rods are prominent. On the rational surfaces, conical process arises from the point of entrance of the optic nerve.

Phono–receptors (Auditory Organ)

The ears are the auditory organs. The external ear is absent, however the middle ear and the internal ear are developed. The tympanic membrane is delicate, small, and circular in shape. The *columella* is rod–shaped. The *internal ear* comprises three *semicircular canals* and a *membranous labyrinth*. The structure in general resembles that of the amphibian ears.

Urino–genital Organs

In chelonians, the excretory and reproductive systems are very closely associated, and as such, they are discussed under urinogenital system. The excretory system comprises the *kidneys, ureters* and the *urinary bladder*. The kidneys are of *metanephric* type confined to the pelvic region. They have lobulated surface and the posterior part is slightly narrow. The ureters arise from the ventral surface of the kidneys. There are short and thin and open into the cloaca. The urinary bladder is a sac–like structure. In some turtles the *accessory urinary bladders* (one pair) are also present which are connected to the cloaca. They perform the function of respiration.

The sexes are separate in chelonians. The *sexual dimorphism* is not very distinct. In some aquatic forms the *musk glands* are present beneath the lower jaw and at the junction of carapace and plastron. The secretion of the musk glands attrack the opposite sex during the breeding season.

The *male reproductive organs* consist of a pair of testes, a pair of vasa deferentia and an intromittant organ (penis). The testis is an oval structure and it is attached to the body wall by a fold of peirtoneum, the *mesorchium*. Both the testes are composed of convoluted *seminiferous tubules*. From each tubule is attached the *vas efferens*. It is a small duct. Several *vasa efferentia* unite to from the *vas deferens*. Each testis gives rise to a vas deferens. Its anterior end is known as *epididymis* which is a convoluted structure. The remaining straight tube opens into a the cloaca posteriorly near the base of the penis. There is an erectile penis. The main erectile tissue is known as the *corpus spongiosum* which contains blood capillaries and nerves. The penis is formed from the floor of the cloaca. There is a groove on the surface serving as passage for sperms. This groove is called the *urethral groove*. It terminates into a *glans*.

The *female reproductive organs* consist of one pair of *ovaries* and one pair of *oviducts*. The ovaries are large and they are situated in the posterior part of the abdominal cavity. Large number of ova are present in the ovary and they are in the different stages of development. The ovaries are attached with the folds of Peritoneum, the *mesovarium*. The oviducts pass along the posterior border of ovaries. The anterior opening of the oviduct is known as *ostium*. The oviduct is tube like structure and it is coiled. Its wall secretes albumin. They become thick walled and form the *uteri* which act as shell gland. Each uterus opens into the cloaca. The ventral wall of the cloaca forms the *clitoris*. The fertilization is internal. The fertilized eggs are laid by the females. The young ones emerge from the eggs.

Biting Mechanism in Snakes

There are altogether more than 2600 species of snakes all over the world including the marine forms, and, out of which only 300 species are poisonous. Number of species diminishes progressively through the polar regions. In India, there are 330 species only and among them only 69 species are poisonous. As per record of the W.H.O., nearly 30000 to 40000 persons die of snake bite in the world every year.

Biting mechanism : The biting apparatus taking part in the biting process are —

1. Poison glands
2. Poison ducts, and
3. Poison teeth or Fangs.

Now, they are being described below in detail :

There is one pair of poison glands each one is situated on either side of the upper jaw. The poison glands are actually the *parotid glands*. Each poison gland is sac like in appearance. They are held in position by some ligaments. With the help of anterior ligament, the gland is attached with the maxilla. The posterior ligament is present between the gland and the quadrate. In addition to these, fan–shaped ligaments are also situated between the side walls and squamoso–quadrate junction.

Each poison gland is provided with a narrow duct at its anterior portion which passes along the side of the upper jaw, loops over itself and finally opens at the base of the fang.

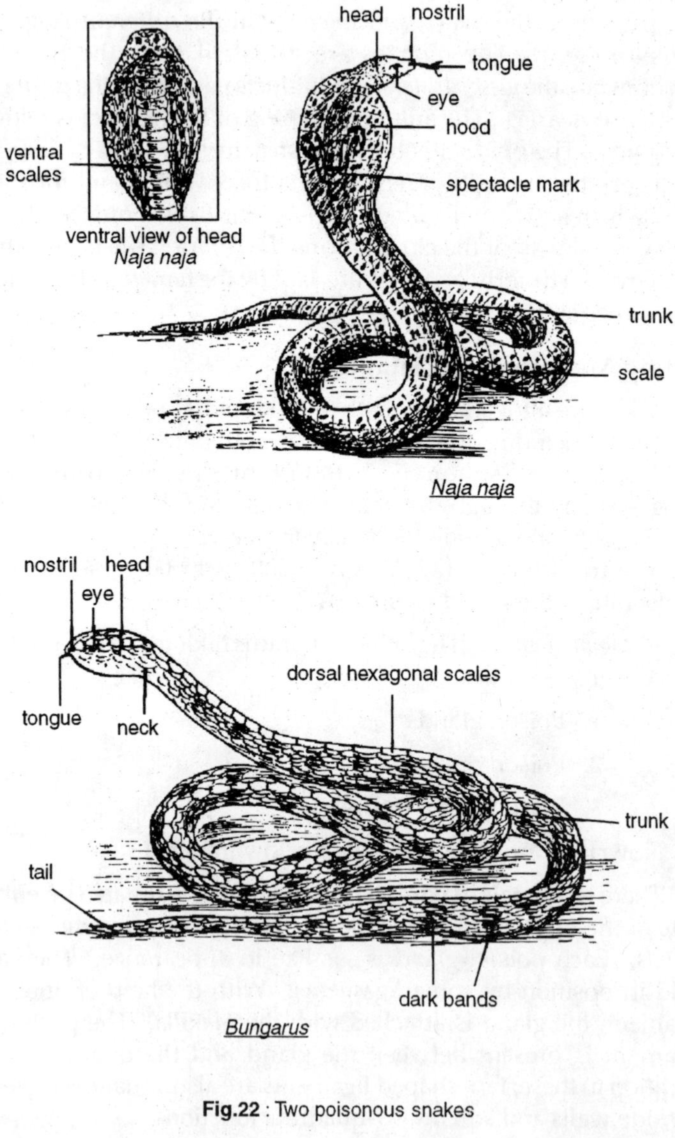

Fig.22 : Two poisonous snakes

There is one pair of fangs in the upper jaw. They are enlarged maxillary teeth which are very sharp and pointed. There is great power of regeneration (when lost for some reason). On the basis of structure and position, the fangs are of the following types :

1) *Proteroglyphous type*: The fangs are comparatively small and they are present in front of the maxillae. The fang has a groove all along its anterior face.

   Examples : Cobra, Krait, Sea snakes and Coral snakes.

2) *Stenoglyphous type*: The fangs are movable and turned inside. Poison canal runs through the fang and opens at the tip.

   Examples : Vipers and Rattle snakes.

3) *Opisthoglyphous type*: The fangs are small and lie at the back portion of maxillae. The fang has a groove along its posterior face.

   Examples : Some colubrid snake (African tree snakes)

4) *Aglyphous type*: Aglyphous dentition is present in the non–poisonous snakes.

Associated bones and muscles

There are some important bones and muscles which are directly or indirectly associated with the mechanism of biting. In the skull, *maxillae, quadrate, pterygoid, squamosals, ectopterygoids* and *palatines* are movably articulated. Premaxillae are very much reduced. Sqamosals are loosely attached to cranium. The joint of quadrate and lower jaws acts as fulcrum. Quadrates are also loosely articulated with the cranium, pterygoid and lower jaw. Ectopterygoid is a transverse bone.

The important muscles are *Digastric muscle, Anterior* and *Posterior temoralis muscles* and *Protractor–Pterygoid* or *Sphenopterygoid muscle*. In addition to these, there are two more muscles associated with the poison glands. These are *masseter muscle* and *Mandibular constrictor* muscle.

The gastric muscle is attached with the squamosal bone anteriorly and with the base of the lower jaw (articular) posteriorly. The Sphenopterygoid muscle is attached to the Sphenoidal region anteriorly and dorsal surface of the Pterygoid posteriorly. Anterior and Posterior temporalis muscles are attached to the side walls of the cranium and the lower jaw.

Opening and Closing of Mouth (*Process of biting*)

(i) When the digastric muscle contracts, the mandible is lowered and the skull alongwith the upper jaws goes up. As a result, the mouth opens.

(ii) The distal end of the quadrate is pushed forward which thrusts the pterygoid, palatine and transverse bar.

(iii) Contraction of the Sphenopterygoid muscle also contributes to the above process and Pterygoid is pulled forward and ectopterygoid is pushed upward.

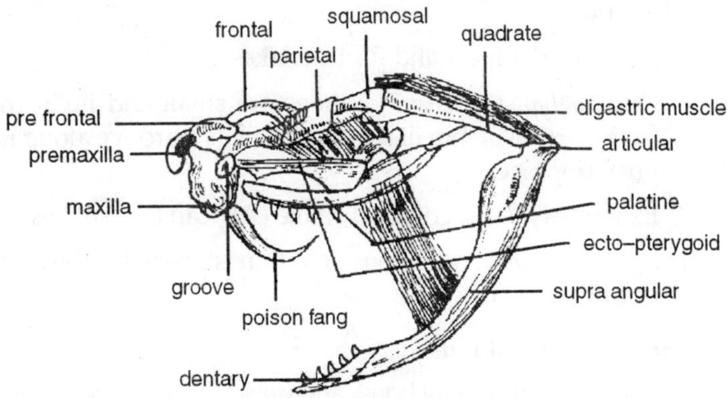

Skull of poisonous snake with jaws partly opened

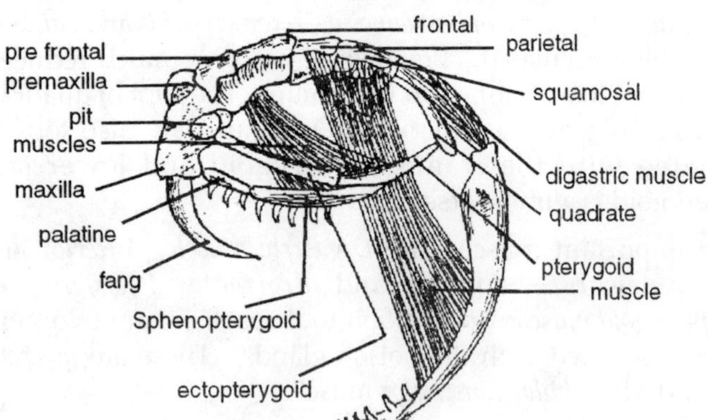

Skull of poisonous snake with fully expanded jaw

Fig.23 : Biting Mechanism of Snakes

(iv) The upward movement of ectopterygoid brings about a rotation of maxilla on its own axis and as a result fangs are erected.

(v) The mouth closes by the contraction of anterior temporalis and pterygoid muscles. Fangs pierce into the skin of the victim.

(vi) Muscles associated with the poison gland (*masseter* and *mandibular*) contract and the poison is squeezed into the body of the victim through the poison ducts and fangs.

Snake Poison (Venom)

Snake venom is complicated mixture of many organic compounds, for example — *Proteolysins, Cardiotoxins, Haemorrhagin, Neurotoxins* and *Antibactericidum* etc.

Various symptoms are shown by the victim of snake bite by different snakes. In the cobra bite the victim feels pain, weakness and difficult breathing. There is profuse salivation and frequent vomiting. The victim also becomes much lethargic. There is nervous breakdown also. In Krait venom the victim feels pain in abdomen, rest symptoms are like cobra. In the case of viper poison, swelling at the place of biting is very common, also rupture of endothelium, hemorrhage and blood clotting are very common signs. Low blood pressure and heart failure are also common.

Some Common Poisonous Snakes

1. *Naja tripudians*
2. *Naja bungarus* (*Naja naja*)
3. *Bungarus fasciatus*
4. *Bungarus niger*
5. *Bungarus candidus*
6. *Ancistrodon himalayanus*
7. *A. hypnale*
8. *Lachesis macrolepis*
9. *Trimeresurus gramineus*
10. *Vipera russelli*
11. *Echis carinatus*
12. *Hydrophis*

13. *Hydrulus platurus*
14. *Callophis macclellandi*
15. *Hemibungarus nigrescens* etc.

Temporal Vacuities of the Skull

There is great diversity in the construction of skull among the different reptilian groups. The composition of the skull in reptiles is somewhat complicated, however, they are of much significance in classification and in tracing the phylogeny of different groups. The skull of *cotylosaur,* the stem reptile, is considered as a morphological basic type. The skull of *Seymouria* however, is supposed to be more primitive standing closer to amphibians. There are several structural variations particularly in the temporal region of the skull with regard to presence or absence of *vacuities* or *fossae* and their modifications in different groups. The common features of the reptilian skull are —

1) *Monocondylic* skull i.e. there is a single occipital condyle.
2) Presence of Parietal foramen for pineal body.

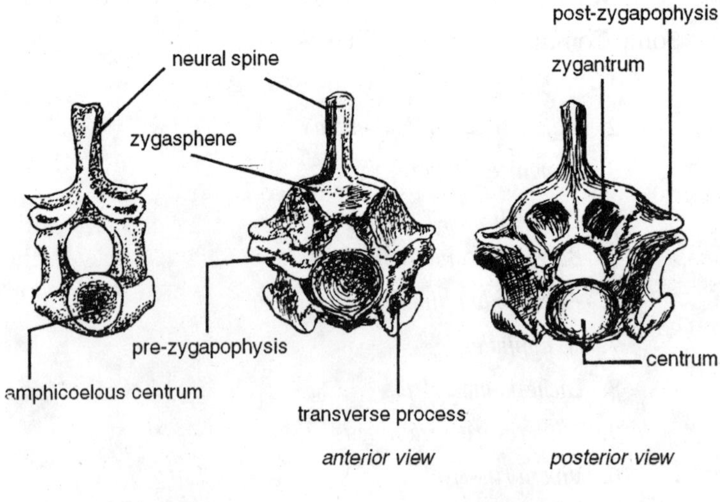

Fig.24 : Vertebrae of Python and Sphenodon

3) Presence of epipterygoid and ectopterygoid bones.
4) Almost complete calcification of the skull except with a very little cartilaginous element in the ethmoid region.
5) Inter–orbtial septum or orbitosphenoid bone present.
6) The skull is of *tropibasic* type i.e. trabeculae meet just in front of the hypophysis.
7) Jaw suspension is streptostylic (i.e. quadrate is movable). or *Monimostylic* (i.e. quadrate is firmly united to squamosal and sometimes other bones).

Temporal Vacuities and Types of Skull

1. *Anapsid type of Skull* : This is the most primitive type of skull. In this case, there is lack of temporal openings, and the skull is complete. This condition is known as Anapsid condition of the skull. For the reason of possessing this type of skull, the orders – Cotylosauria and Chelonia have been grouped together in sub–class – *Anapsida*.

   In more advanced reptiles temporal openings are met with in the following patterns, on each side of the skull.

2. *Synapsid type of Skull* : In this case, a single opening is bounded laterally by an arch composed mostly of Jugal bone. The vacuity is inferior in position surrounded by Posts–arbital, squamosal and jugal. Parietal bone does not take part in the formation of temporal vacuity. The synapsid condition of the skull occurs in *Plecosaurs* and *Therapsids* and they come under the sub-class – *Synapsida*.
3. *Parapsid type of Skull* : In this case also, there is single opening which is located dorsally and is superior in position. The vacuity is surrounded by parietal, post–orbital and squamosal. Jugal does not take part in the formation of the fossa. This arrangement has been found in *Ichthyosaurs* and *Plesiosaurs* and they come under the sub–class – *Parapsida*.
4. *Diapsid type of Skull* : In this case, two temporal vacuities are present in the dorsal side. Post–orbital and squamosal mainly contribute in the formation of vacuities. This condition is very common in *Lepidosauria* and *Archosauria*. But, some sort of modifications may occur in the different animals belonging to sub–class – *Diapsida* (having diapsid

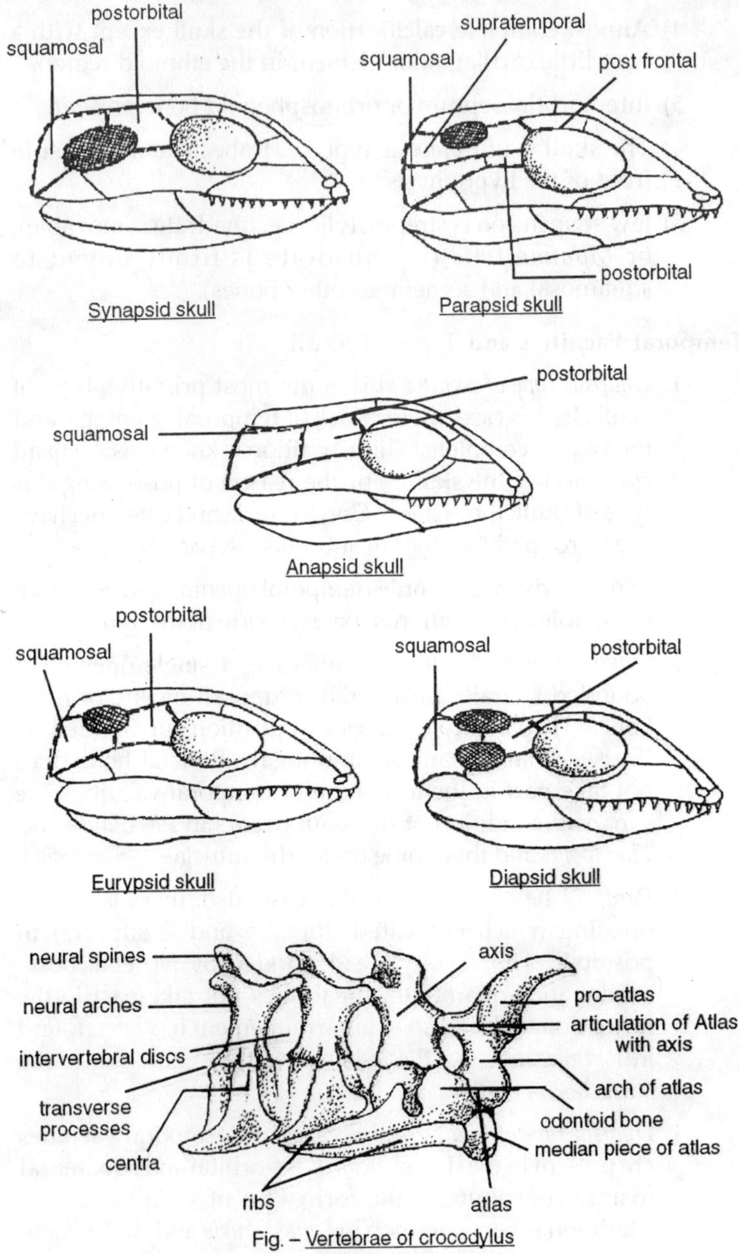

Fig.25 : Basic structure of reptilian skull, and vertebrae of *crocodylus*

skull). In lizard, quadrato–jugal is lost, thus there is reduction of marginal bones at the lower temporal opening which is no longer enclosed. Thus, the skull does not look like a diapsid skull. In snakes, the margin of the upper opening is lost. Post–orbital and jugal are reduced. The reduction is correlated with the release of the quadrate and development of streptostyly (moveable quadrate).

5. *Eurypsid type of Skull* (*Eurypsida*) : In this case a supratemporal fossa is present. It is bounded medially by the parietal and bordered by the post–orbital and squamosal. This condition is present in *Protosaurus* and *Sauropterygians*.

Chapter 6

Class — Aves

All the birds belong to class – Aves. They are warm–blooded (endothermal) bipedal creatures. The body is covered with feathers which are epidermal exoskeleton. Beak and claws are covered with horny sheaths and feet are covered with minute scales. There is a single occipital condyle. Forelimbs are modified to form wings with a *carpo–metacarpus* and the first three digits of the hand. Hind limbs support the body weight when the bird is standing. There is a *tibiotarsus* and a *tarsometatarsus*. Bones are light and have air spaces and *pneumatic foramen*. The cervical and thoracic vertebrae are *heterocoelous* and the ribs possess *uncinate process*. Several vertebrae fuse to form a *synsacrum*. The Sternum has a ventral keel. Furcula bone is formed by the union of clavicle and interclavicle. *Arytenoid* are frequently ossified in birds. The ilium is very large and pubis and ischium are directed backwards. A crop is present in the alimentary canal and the stomach is divided into a gizzard and a proventriculus. The larynx is rudimentary and its place is taken by *syrinx*. *Air sacs* are attached with the lungs. The heart is four chambered and only the right aortic arch is present. There is a oil gland near the tail portion. Optic lobes are well developed and they are lateral in position. In the lumbar region of the spinal cord there is a *sinus rhomboidalis*. A large *pecten* is present in the eye. The kidney is *metanephric* and there is no urinary bladder. Right ovary and oviducts are atrophied. They are *oviparous* with yolky eggs.

Origin and Evolution of Birds

The chief glory of the reptiles lies in the fact that they gave rise to warmblooded feathered birds and warm-blooded hairy mammals. Undoubtedly, these two groups of animals are ruling the world and they have evolved from the reptiles. There are many features which are common to both reptiles and birds and as such, some authors have called the birds as *"feathered reptiles"*. Huxley called them *"Glorified reptiles"* and placed them in the group *sauropsida*. As a matter of fact, the birds are maintaining the glory of the past reptilian age, of course not as a reptile, but as a bird, not on the earth but in the air mostly. In addition to features that are common to both birds and reptiles, birds have definitely developed some special features for aerial adaptation.

Earliest (fossil) Birds

There is great scarcity of birds' fossils. The earliest recorded birds are known from the two well preserved specimens from Jurassic of Bavaria. They are : (I) *Archaeopteryx lithographica* and (ii) *Archaeornis siemensi*. They are supposed to be connecting link between reptiles and birds. Had there been no features, they would have been conveniently classified as reptiles. *Archaeopteryx* possessed the following characters —

1) The body axis was elongated and lizard like.
2) Fore limbs modified as wings and body covered with feathers
3) Absence of bill
4) Presence of teeth on both the jaws
5) Simple amphicoclous vertebrae
6) A fixed rather than free quadrate
7) Interclavicle present
8) Lack of differentiation between cervical and other vertebrae
9) Sacral vertebrae few in number (5-6) and cervical vertebrae 9 or 10
10) Ilium elongated and pubis directed backwards
11) Public symphysis present unlike birds
12) A long tail with 20 vertebrae
13) Presence of ventral abdominal ribs
14) Humerus longer
15) Sclerotic bone present round the eye ball
16) Bones heavy, not pneumatic

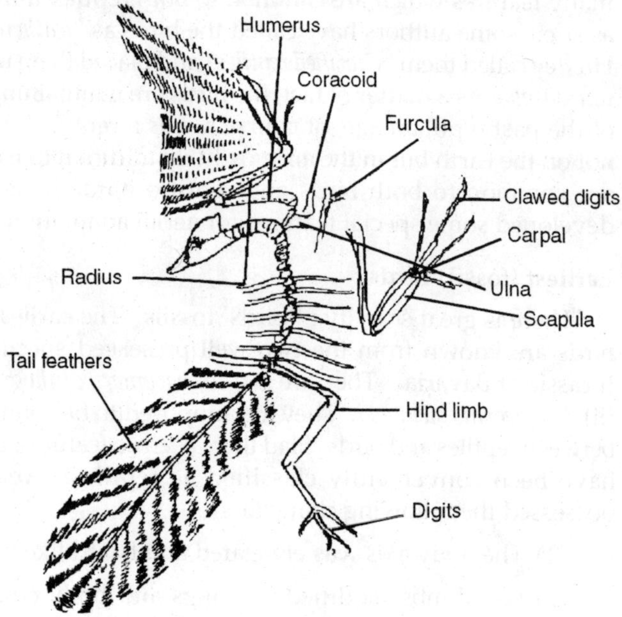

Fig. – Skeleton of *Archaeopteryx* (Fossil form)

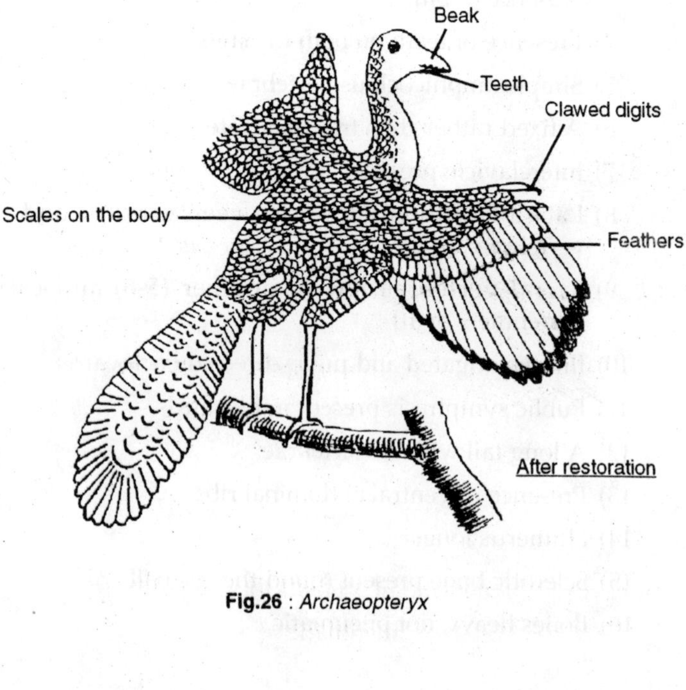

Fig.26 : *Archaeopteryx*

17) Fingers with claws
18) Weak and unossified sternum
19) A short ilium present
20) Fibula is long
21) Carpometacarpus absent and separate metatarsals
22) Absence of uncinate process of the ribs
23) Some scales are also present on body and limbs

Thus, the fossil birds possessed more reptilian than avian characters. The most important avian characters are the presence of feathers and feet. Heilman has designated them as "warmblooded reptiles disguised as birds".

Analysis of the Characters of Modern Birds

Even the present day birds represent many feathres whicn are in favour of the view that the birds originated from the reptiles. The main point of resemblances are the following —

1) Diapsid condition of the skull found in Archosauria
2) Monocondylic skull in both the groups
3) The middle ear has a single delicate rod–shaped columella
4) Presence of 6 bones in each mandibular ramus
5) Epiphysis of the vertebrae absent
6) A precoracoid bone is present in Rhea also
7) The uncinate process of the ribs are present in *Sphenodon* also
8) The bipedal condition was present in dinosaurs also
9) The pecten well developed in birds is also found in a rudimentary condition in crocodiles
10) The heart in birds is four chambered as in crocodile but the left aortic arch is abolished in birds
11) The ankle–joints between proximal and distal tarsals resemble those of reptiles
12) Air sacs present in birds can also be seen among chamaeleones
13) A claw on thumb of a bird wing shows relic of reptilian ancestry

14) Presence of scales and claws on legs
15) Presence of sclerotic plates in the eye, also found in fish like reptile, Ichthyosaurs
16) Metanephric type of kidneys in both the groups
17) Tetraradiate type of pelvis as in *ornithischian dinosaurs* (bird like reptiles)
18) Telolecithal egg and meroblastic cleavage in both reptiles and birds
19) Extra embryonic membranes present in both the groups during development
20) Hatching of young ones similar in both groups
21) The brain of birds seems to be an elaboration of the brain of reptiles.

Thus, from the study of fossil and modern birds it becomes evident that the birds have originated from the reptilian stock. Sufficient specialisation and adaptational changes have taken place in the modern birds for leading an aerial life. So, the statement that *'Birds are glorified reptiles'* as made by Huxley becomes established.

Origin of Flight

Who were the ancestors of birds, is an open question. How flight was initiated and the birds were separated from their ancestor is intimately connected with the origin of flight. Several theories accounting for the origin of flight have been discussed below which were proposed from time to time.

(1) *Cursorial Theory of the Origin of Flight*: This theory has been advocated by *Nopsca*, a Hungarian Palaeontologist. According to his view, the acquisition of the power of flight was the assumption of a semi–erect posture. The legs were main organs of locomotion. It was presumed that the ancestor was a slender hollow–lobed bipedal creature which ran as readily as it walked. Continued use of arms as propellers brought about constant increase in the size of the scales on them. After a very long time, when the fore arms were constantly used, the scales changes into feathers. Broad arms contributed much in increasing the speed of the animals. After many generations, power of

short flight first and then long flight was attained. The first bird evolved from some long tailed reptiles, a hypothetical 'pro–avis'. Thus the feather is regarded as the specialised scales.

But, the fact that feathers evolved first on the wings and then spread all over the body is a matter to be criticized. It has been shown that the scales and feathers are different structures arising from the different layers of cell in the skin. Moreover, there is nothing to suggest that the fossil birds *Archaeopteryx* and *Archaeornis* were rapid runners. So, this theory has not been accepted by all.

(2) *Arboreal Theory of the Origin of Flight*: This theory was proposed by *William Beebe*. There are two phases of this theory —

(i) The pair–wing theory

(ii) The four–wing theory or Tetrapteryx theory

The pair–wing theory is based on the characters of *Archaeopteryx* mainly. But, small expanse of wings, weak sternum and lack of strong flight muscles are such characters which show no possibility that they could be strong fliers. Naturally, propulsion of wings was not necessary. They were soarers or gliders and glided from one tree to another in the manner of flying squirrels. Moreover, long clawed legs and prehensile tail point out toward an arboreal habitat, thus supporting the arboreal origin of flight.

According to the four–wing theory of Beebe, the ancestors of the birds were arboreal reptiles which were in the habit of jumping from branch to branch and from one tree to another. The individuals with larger scales on arms and legs were able to make better swooping jumps. After some generations, they became better gliders, beating the air with fore and hind limbs. The scales were gradually transformed into feathers by the fraying of the edges. Later on, the birds discovered that they could do better by using only the anterior pair and as such the posterior pair became inactive and after many generations the feathers of the hind limbs were lost.

The author claimed that even in some modern birds the vestigial flight feathers occur on the thighs. Similar feathers were present on the thigh of Archaeopteryx. The main idea of this theory is that the birds were scansorial arboreal reptiles. But there is no evidence that the birds ever possessed four wings and thus, the theory has been disapproved.

(3) *Compromise Theory of the Origin of Flight*: This theory was advocated by W.K. Gregory. The author believed in double origin i.e. some from cursorial and others from arboreal ancestors.

According to this author, the ancestors were the scansorial (climbers) arboreal reptiles and were capable of running on the ground and also they could jump from tree to tree. They acquired the habit of perching upright on the branches. Grasping first digit and strong claws of *Archaeopteryx* added to the authenticity of this hypothesis. This theory is of course more acceptable than the two theories discussed above.

(4) *Gliding Theory of the Origin of Flight*: The gliding theory of the origin of flight was proposed by Heilmann. It is based on the fundamental ideas of Beebe and Gregory with some additions of his own. According to him the birds and certain dinosaurs came from a common arboreal ancestral stock. The dinosaurs early returned to a terrestrial life but in the case of birds the return was much delayed until the assumption of flight.

(5) *Diving Theory of the Origin of Flight*: This theory was proposed by Newman (1939). This theory advocates that the flight originated in connection with soaring over the water and diving after the manner of a fish. Newman held the view that pro–aves used the forelimbs together with their membranes and elongated scales or plumes which could be of great use in diving. These accessories helped in making a choice of direction and by more flapping of the wings the pro–aves could perform longer glide over the water. Thus circling movement was easier during search for fish. The pro–wings served useful purpose in running or leaping from bough to bough.

It was supposed that the *Penguins*, the most completely marine of all birds are able to swim underwater fastly. Webbed toes and wings were reduced to form strong flippers. These birds under went a specialization of the primitive wings, using them for under water flying. *Grebes* never developed fully effective organs of flight, but *loons* became good fliers through its diving propensity has been retained.

The strong flying birds are the derivatives of the ancestral diving types which gradually perfected their flight, while the land birds of all sorts have been derived from the sea birds. There are some evidences that sea is the ancestral home of birds and they invaded the land later on.

Origin and Ancestry of Birds

Regarding the ancestry of birds three theories have been proposed from time to time. The theories advanced have been discussed below in detail :

1 From *Pterosaurs* (Flying reptiles) : According to this proposition, the flying birds have taken their origin from the archosaurian–diapsid, the *Pterosaurs* (flying reptiles). The resemblances extend to the following points —

(i) In both the groups, conical teeth were present and in the birds they were replaced by bill for quick collection of the food materials.

(ii) The skull is held almost at right angle to the back bone.

(iii) Nearly similar brain structure in both groups

(iv) Sclerotic plate and large orbit present

(v) Large sternum in both the cases which were not properly ossified.

(vi) Some bones were of pneumatic nature in both groups

(vii) Bipedal condition was common in both.

But, there were some valid objections also. The contrasting wing structure in two groups clearly indicates that they evolved independently along different lines. Birds are more like *saurischian dinosaurs* than *ornithischian dinosaurs* in the structure of their pelvis. In dinosaurs there was no clavicle and there were some other skeletal differences too. So, the origin of birds from *Pteropus* has not received general acceptance. The fling habit of *Pterosaurs* is only a matter of convergence.

2. From *Bipedal dinosaurs* : The idea that the birds have taken its origin from the *bipedal dinosaurs* has widely been accepted. The features which are common to both the birds and the bipedal dinosaurs have been summarised below —

    (i) In *theropodan dinosaurs,* the bones were hollow.

    (ii) The pelvis in *ornithischia* was shaped like that of birds. Ischium and pubis are greatly prolonged, Ilium greatly expanded thinned. The acetabulum is partly open. In ornithischia, a large *"prepubic"* process corresponding to the *"pectineal process"* is present.

    (iii) The ankle–joints between two rows of tarsals

    (iv) The pelvis of some dinosaurs is large and upto 8 vertebrae are present in the sacrum

    (v) In some *therapodans,* the hind feet possessed only four clawed toes (the fifth being absent) and the first one rotated behind the metatarsals like the birds.

    But in the opinions of many, it does not seem convincing that the birds arose from the large heavy–boned ornithischians. There is no prepubis in bird. Also, no member of *therapod* was an arboreal creature showing tendency of assumption of wings from the forelimbs. There is no relationship in the matter of possession of clavicles.

3. From *Pseudosuchian* reptiles : It is believed that the Dinosaurs and the birds have originated from a common ancestral stock. Birds resemble with the *Archosaurian diapsids* and particularly with the *pseudosuchian reptiles* (Thecodontia) of Triassic period. They are considered as the ancestor of the *Theropodan dinosaurs* and also the ancestor of the birds. The structure of the skull of *Archaeornis* forms the basis of this concept. The skull shows —

    (i) That the quadrate is immovably articulated with the squamosal.

    (ii) Short quadrato–jugal and ascending process of jugal behind the orbit.

    (iii) The skull of *archaeornis* is similar to that of the *Pseudosuchian Euparkeria*. The only differences are —

1) The dorsal temporal opening of the skull of *Euparkeria* is absent in *Archaeornis*.
2) The parietals, post orbitals and squamosals of the Archaeornis grow and obliterate the upper fenestra.

Some avian characters of *Euparkeria* as mentioned by Heilmann and Broom are —

1) Bipedal condition
2) Forelimbs shorter than the hindlimbs
3) Bird like tongue
4) Ventral ribs present
5) Double headed ribs and uncinate process small
6) Clavicles long and slender
7) Pelvis resembles the avian pelvis of the embryonic period
8) Cervical and pre–sacral vertebrae almost similar in number

Conclusion

On the basis of various characters analysed above, it may be conclusively said that birds arose from *archosaurian diapsid reptiles* possibly from *Pseudosuchian bipedal stock* (Thecodontia). Thus, the *Pseudosuchian Euparkeria* gave rise to birds as an independent offshoot and also to the reptiles like dinosaurs etc. in other independent line. *Archaeopteryx* is the earliest bird (fossil) at present known. But, there is possibility of several stages between the reptiles and the Archaeopteryx which could not be revealed by the geological records.

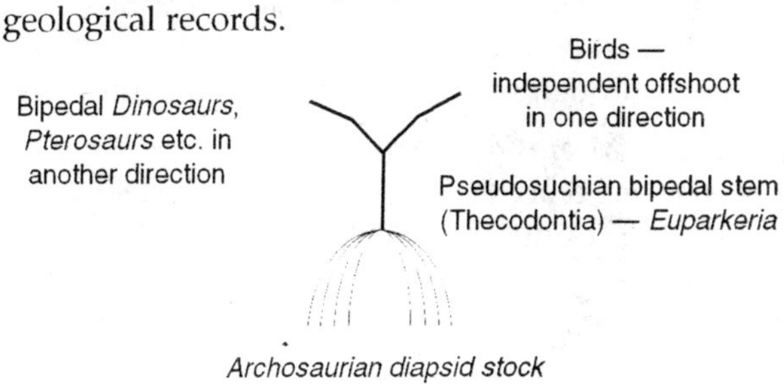

Migration of Birds

Bird migration is a fascinating phenomenon which has provoked more speculation than any other aspect of Avian biology. It is a periodic movement made necessary mainly for food requirement and for reproductive purposes. It is a striking regularity year by year in the arrival and departure of migrants. Migration is a dynamic response in every sense of the term. All the birds do not migrate, they are known as *residents* and those birds who necessarily migrate with regularity, are known as *migratory birds*. Migration has been observed not only in the terrestrial birds, but, the aquatic birds also migrate. The desire of migration is a hereditary impulse. Our knowledge of migration of birds is still far from complete. Various aspects of the bird migration has been discussed below.

Types of Migration

(1) *Latitudinal Migration* : That the birds fly from north to south or vice–versa, linking the equatorial regions of each hemisphere with its temperate and cold areas, is an old concept. However, it is correct that latitudinal migration takes place for change of climatic conditions. Latitudinal migration can be seen in several birds e.g. Golden plover (*Pluvialis dominica*) of America which spends nine months of winter, 8000 miles south in Argentina.

(2) *Altitudinal Migration* : It is also known as vertical migration. Birds reach the mountain regions in the summer season and return to the plains during winder season. High elevation birds because of excessive cold are rather forced to come to the ground. In the spring, they again migrate to higher altitudes e.g. many hill birds (e.g. *Scolopax rusticola*).

(3) *Longitudinal Migration* : This sort of migration takes place in the East–West direction. The starling of eastern Europe passes through Atlantic coast to avoid continental water.

(4) *Vagrant Migration* : This migration is irregular and wandering type. In the case of herons and terns the adults and the young ones disperse in all directions after breeding.

(5) a) *Summer visitors* : During the spring season, some birds arrive from the south and stay to breed, and then leave for the south again in the autumn. These birds are known as summer visitors e.g. cuckoo, swallows etc.

b) *Winter visitors* : In the autumn some birds arrive from the north, they stay through out the winter season and with the advent of spring they fly northward again e.g. Redwing (*Turdus musicus*) in Britain.

c) *Birds of Passage* : These birds are seen for a short time twice a year when on their way to colder or warmer countries, e.g. Sand piper (*Tringa hypolecos*). They winter southern hemisphere. European Cranes and American Sanderling are also the birds of passage.

d) *Resident birds* : These birds are more or less static through out the year but they may show partial migration also. The American Red–eyed Vireo is an example of summer resident. Some European species of swallows are also good examples. They always breed in a particular area.

Other Less Important Types of Migration

1. *Partial Migration* : All the birds of a group of migratory birds do not leave the home land. They stay at the breeding place. Such a migration is known as partial migration. e.g. American blue jay (*Cynocitta cristata*).
2. *Local Migration* : Local migration is seen among the resident birds only. The local migratory movements are prominent when the change of season is more pronounced e.g. the paradise birds along the base of the Himalayas.
3. *Abnormal local Migration* : Under stress of abnormal conditions, birds can be driven out from one habitat to another for their survival.

Different Aspects of Migration

1. *Nocturnal Migration* : There are several migratory birds who prefer to travel at night. For example, passerine birds normally travel at night and take rest by day. This habit provides them better protection against birds of prey. During the day time they are engaged in collecting their food.
2. *Diurnal Migration* : There are several birds which prefer to travel during day time. For example, falcons, herons, swallows and humming birds etc. like to travel during the day time. But some birds like ducks, loons and terns etc. can travel both during the day time or at the night irrespective of light or darkness.

3. *Range of Migration including velocity and altitude* : Flight of flocks of geese has been recorded at a height of 8500 feet above the North Sea during the storm but in the normal condition the height is only 5000 feet over a distance of eighty miles. The flight of small passerine birds is very much affected by the force and direction of wind. The altitude is not more than 300 feet on this case. Radar plotting has shown that some geese fly at an average speed of twenty miles per hour. A finch can fly about 370 miles in a day and a Shrike upto 125 miles as per available records. The arctic tern (*Sterna paradisea*) covers a distance of about 22000 miles in going out and coming to the native place during its migration. The golden plover migrates from Hudson Bay and Alaska to South America covering a distance of 2400 miles in 48 hours which is a non–stop flight. The sand piper (*Actitis macularis*) flies at a speed of 25 miles an hour and the Hawks at 30-40 miles per hour. The intensity of light, rain, visibility and other weather conditions affect the velocity of flight.

4. *Methods of Migration* : Some birds migrate in groups but there are some birds which prefer to travel alone. The Golden oriole and many birds of prey travel alone during migration. Most of the passerine birds travel in dense groups without particular formation. However, some birds like cranes, ducks, geese and storks etc. fly in skein, line or V–formation. There is a continuous flight in a single direction. These flight patterns have certain advantages e.g. visibility is increased and all participants are able to see each other and they get encouragement and idea of the flight direction. Several birds reach their destination in a single flight without break of journey but some of the birds take rest in the way at some selected places when the route is longer. Examples have been cited later.

5. *Course of Migration and Migratory routes of some birds* : Direction and distance are the two important features of migratory routes. The migratory birds try to avoid great barriers like oceans, seas, deserts and mountain chains, normally.

   The European migrant bird *Hirundo rustica* migrates from Europe to Africa. It resides in the temperate parts of

Europe from April to September. The home land is its breeding place. It migrates to Sudan, in West Africa and to many parts of Southern Africa even at the cape of Good Hope where it winters from October to onwards.

Many passerine birds of Europe, migrating to Africa avoid long passages over the Mediterranean and prefer shorter route from Spain to Morocco. They also prefer to the routes from Italy and Sicily to Tunisia and Asia Minor to Egypt. In some cases deserts are also crossed and oases are used as resting place. Sahara desert has been used as winter retreat. *Anser indicus* and *Grus grus* cross the Himalayas in both directions when they fly between winter quarters in India. Water birds can cover vast stretches of the oceans during their migrations. The Tahitian or Bristle–thighed curlew (*Numenius tahitiensis*) travels thousands of miles across the Pacific between various islands where it winters. They make their nests at Alaska.

The European Quail (*Coturnix coturnix*) winders on the shores of the Mediterranean and even in Egypt and tropical Africa. It takes very lengthy migration. The breeding area has been extended far to the north only to return every year to the ancestral feeding area.

Terns rest in northern Europe, Asia and Alaska and travel to the Antarctic where they spend the southern summer. They have to fly between 7500 and 9500 miles each way along the sea coasts or over the sea. The Arctic Tern (*Sterna paradisea*) travels with Great shearwater (*Puffinus gravis*). The Tern nests in the northern summer from June to August and the Petrel resides in northern Europe without nesting during that period. Between September and November both migrate to the southern hemisphere. Now, during the southern summer (from December to March) the shearwater nests in southern Africa and the Tern remains the resident. So, this migration becomes interesting. Between April and May both species return to the north.

6. *Regularity of Migratory travels* : Migration takes every year with regularity and accuracy. Several birds visit the same locality every year and they return to their original place

when the purpose of migration is served. They have very highly developed place–memory. They have very keen power orientation and a sense of direction which enable them to recognise the concerned places during migration. At the time of returning to their native places even the young birds accomplish the journey nicely. The young birds have no previous experience of either route or destination. Naturally, it seems to be the hereditary impulse. Sensibility to terrestrial magnetism and visual recognition of land mark help the birds in finding their way. In the spring, the adult males are first to arrive on the breeding grounds. At the time of returning in the autumn, the young birds lead. Birds can grasp some idea of the earth's movement which might help in locating various places on the globe. They are very sensitive to intensity of light also. In the spring season the adult males arrive on the breeding ground first and then they are followed by the females.

Causes of Migration

Though the desire to migrate is a hereditary impulse, however, it is influenced by some external and internal causes. The external causes include —

1) The scarcity of food in the original locality where the birds reside permanently
2) Shortening of day length
3) Increase of cold and sudden changes in the weather conditions continued for a longer time

The internal stimulus is the maturation of gonads and secretion of certain concerned hormones. It has been suggested by Rowen (1925) that the birds migrate when the gonads are in changing condition. The secretion of interstitial cells in males might be providing stimulus for migration. Increase or decrease in day light affects the maturation of gonads, and thus, migration is affected.

The pituitary hormones are also concerned with the migration of birds as they stimulate the activities of the gonads in both the male and female individuals. Fat deposition in the migratory birds has also been considered as the controlling factor by some workers. Increasing day length affects the activities of the *hypothalamus,*

Pituitary and gonads, which are functionally linked. Thus, the physiological changes in the body induce readiness to migrate. The thyroid hormones are also said to be directly or indirectly concerned with the migratory activities of the birds.

Advantages of Migration

The following are the main points of advantages during migration period.

(i) Birds avoid excessive cold and stormy weather for their safety.

(ii) Short day light is avoided which is not sufficient and convenient for food collection.

(iii) Adverse conditions are avoided so that, scarcity of food and water can not be faced.

(iv) Suitable uncongested nesting places are made available.

(v) Long hours of day light becomes available

(vi) Birds get an abundance of food supply

Conclusion

The phenomenon of migration is very much essential for the birds. If the suitable breeding area and feeding area are not available to birds at the proper time, several bird species will be subjected to natural elimination. Suitable climatic conditions for safety and comfortable breeding are necessary for them.

Flight Adaptations in Birds

Birds are warmblooded vertebrates specially adapted for flight. They are perfectly suited for an aerial life which becomes evident from their anatomical peculiarities. There are several modifications (external and internal) in various systems of the body which are considered as flight adaptations and these are of great use in leading a volant life. The important features related to different systems have been discussed below :

External Features

1) The body is stream–lined i.e. like the shape of the aeroplane so as to offer least resistance during flight in the air.

2) Fore–limbs are modified as wings and act as propellers.

3) Body is covered with feathers which provide non–conducting and wet–resisting surface. They prevent heat loss from the body surface.
4) The quill feathers (contour feathers) occurring on the wings (Remiges) and on the tail (Rectrices) are of great use in the process of flight. The former (remiges) is concerned with controlling the rate of flight and the latter (rectrices) acts as elevator and depressor.
5) The hind limbs are used when the birds are walking on the ground. They support the weight of the body. They help in perching when the birds are sitting on the trees.
6) The 'oil gland' is present on the surface of the uropygium and the secretion of this gland lubricates the feathers.
7) Birds have very long and flexible neck, and because of the presence of single occipital condyle, the head can be moved in all directions, so the neck is mobile.
8) The eyes are very important and power of vision is much advanced.

Mouth and Alimentary Canal

1. The mouth is bounded by horny upper and lower beaks devoid of teeth. They act like forceps and collect the food material in the shortest possible time. Because of the absence of teeth, the weight of the body is considerably reduced.
2. A crop is present which acts as reservoir for the collected food to be utilized when needed.
3. A gizzard is also present which acts as a grinding machine discharging the function of teeth.
4. Rectum is much reduced and so the faecal matter does not accumulate in the body to increase its weight.

Muscular System

There are several muscles which are concerned with flight. Important among them are —

1. *Pectoralis major* : It is a large muscle attached with the keel of the sternum. It takes part in causing downward stroke of the wings. This muscle arises from the ventral side of the sternal keel.

2. *Pectoralis minor* : This muscle is also attached with the keel of the sternum and is overlapped by Pectoralis major muscle. It takes part in causing upward stroke of the wings. It arises from the anterior part of the sternum.

3. *Coracobrachialis* : *Coracobrachialis brevis* and *Coracobrachialis longus* are attached to the coracoid and supplement the action of pectoralis major. These muscles arise from coracoid and scapula and are joined to the head of the humerus bone. The help in rotating the wing in the glenoid cavity.

4. *Tensor muscles* : There are three tensor muscles which keep the pre–patagium tensely stretched. These muscles are *tensor longus, tensor brevis* and *tensor accessorius*. A similar tensor muscle, known as *tensor posterioris* keeps the post patagium tensed.

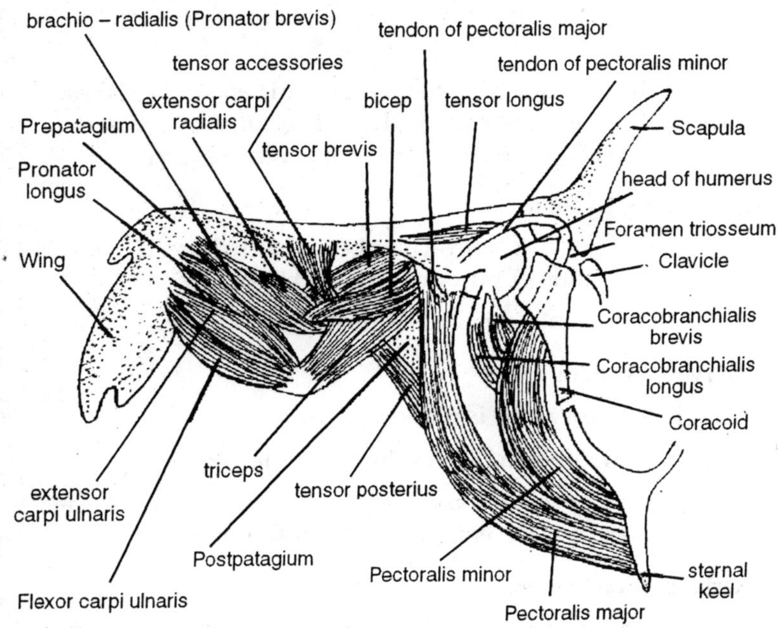

Fig.27 : Flight muscles of Pigeon

In addition to the above described four muscles, these are several accessory muscles e.g. *biceps, triceps, extensor carpiradialis, extensor carpi ulnaris, brachioradialis, digital* and *policis*.

Besides these muscles, there are some muscles concerned with perching mechanism in birds. They include two sets of muscles—

1. Flexor muscles – a) *Ambiens muscle*
   b) *Peroneus medius muscle*
   c) *Gastrocnemius muscle*
   d) *Flexor Perforans muscle*
2. Extensor muscles

Respiratory System

1. *Syrinx* is well developed in singing birds.
2. There is one pair of lungs and attached to them are nine thin–walled air–sacs which increase the volume of respiratory apparatus. Air sacs hold a very large reserve of air.

   The respiratory system is very efficient and large supply of energy is liberated. There is great production of heat in the body, necessary for upper atmosphere.

Circulatory System

1. The heart is four–chambered and there is complete separation of pure and impure blood.
2. Only right aortic arch is present.

Nervous System

1. Olfactory lobe is reduced
2. Optic lobes and cerebellum well developed
3. Eyes specially modified for distinct vision. Pecten is present at the entrance of the optic nerve.

Urinogenital System

1. The urinary bladder is absent which is a device to reduce the weight of the body.
2. A single left ovary is present in the adult female – thus the body weight is reduced.

3. The cloaca is three chambered –
   a) Coprodaeum – it receives rectum.
   b) Urodaeum – where the urinary and genital ducts open.
   c) Proctodaeum – opens externally by anus.
4. Eggs are protected with shell and they are yolky. The fertilised eggs are regularly incubated.

Skeletal System

The skeletal modifications of the birds are associated with their flying habit and bipedal progression. The whole skeletal system is so built that it provides rigidity, lightness and maximum mechanical strength. Fusion of several bones makes the body compact. Most of the bones are of *pneumatic* characters. The important features of the skeletal system have been mentioned below :

1. *Skull*

   (i) The skull bones are very thin and they fuse to form a capsule of bones. The skull is light and protective. The heavy teeth have been replaced by light beak.

   (ii) There is a single occipital condyle (Monocondylic)

   (iii) The Zygomatic arch is reduced, post frontal disappears and the palatines and pterygoids are greatly narrowed.

2. *Vertebrae*

   (i) The heterocoelous vertebrae provide great flexibility particularly for the movement of the neck. The neck can reach every part of the body with its beak.

   (ii) The thoracic vertebrae are fused and support the weight of the body during flight.

   (iii) The ribs of the thoracic vertebrae are flattened and they provide largest surface for attachment of muscles. Sternal parts of the ribs are bony and make a firm thoracic basket to which the flight muscles are attached. The *uncinate process* of the ribs helps in making the thoracic basket compact.

   (iv) There is well developed keel for the attachment of pectoral muscles.

(v) The last thoracic, all lumber, the two sacral and the anterior 5 caudal vertebrae fuse to form a solid *Synsacrum* which supports the pelvic girdle which

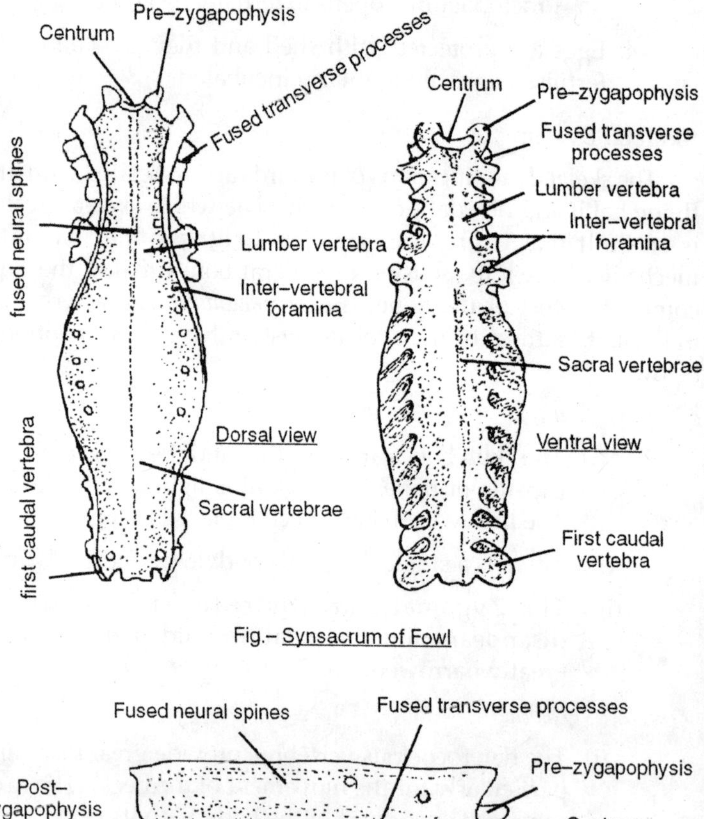

Fig.– Synsacrum of Fowl

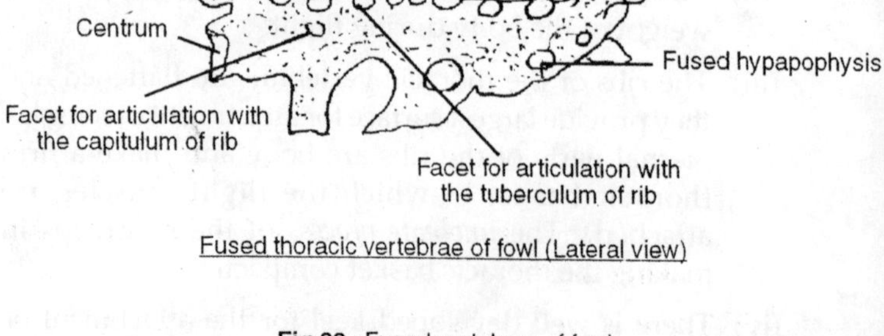

Fused thoracic vertebrae of fowl (Lateral view)

Fig.28 : Fused vertebrae of fowl

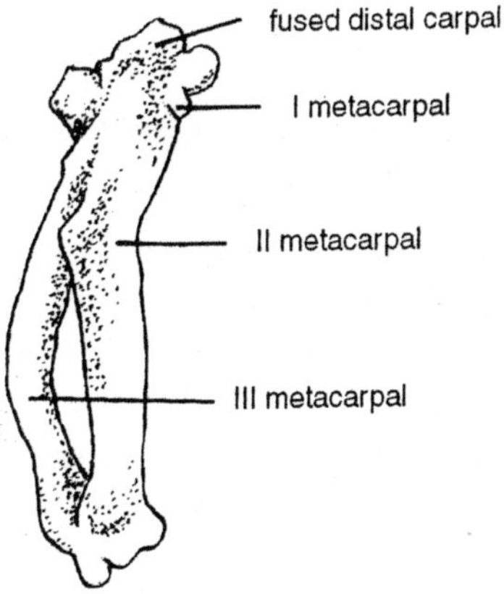

Fig.– Carpometacarpus of fowl

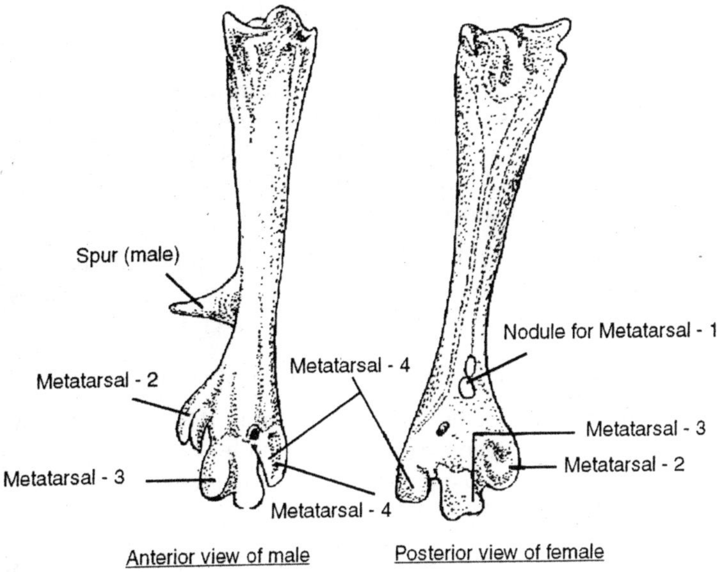

Fig.– *Tarsometatarsus* of fowl

Fig.29 : Fused vertebrae of fowl

bears the weight of the body when the bird walks on the ground.

(vi) The caudal vertebrae are much reduced and form the *pygostyle* which assists stability in air.

3. *Limbs and Girdles*

(i) Humerus is large and possesses a prominent ridge for attachment of muscles. There is a pneumatic foramen also. Carpals and metacarpals are fused to form *Carpo–metacarpus.*

(ii) The fusion of distal tarsals with the metatarsals forms a *tarsometatarsus.* Also, the proximal tarsals fuse with the lower end of the tibia to form *tibiotarsus.*

(iii) Arched clavicles and strong pillar like coracoid of pectoral girdle are situated to resist the inward pressure of the down stroke of the wings.

(iv) Clavicles are ventrally united to form a U–shaped *furcula* bone.

(v) The pelvic girdle is elongated in the direction of the long axis of the body. The fusion of the pelvis with Synsacrum supports the weight of the body when the birds walk on the ground, and also counteracts the effect of shocks.

Thus it becomes evident that the birds are fully equipped for terrestrial, arboreal and aerial environments. In all systems of the body there are some modifications which are adaptations for flight.

Respiratory System in Birds (*Columba livia*, as a typical case)

The respiratory system of birds is highly developed as they require sufficient oxygen for their flight activities. The respiratory system comprises the *respiratory tract*, the *lungs* and the *air sacs*. They are being described below :

Respiratory tract : The respiratory tract includes the nostrils, nasal sacs, glottis, rudimentary larynx, trachea and the Syrinx.

There is one pair of slit like openings at the base of the upper beak which are known as the *nostrils* or *nares*. The nostrils are overlapped by *cere*. The external nostrils open into the *nasal sacs* which communicate to the pharynx by internal nares. The glottis

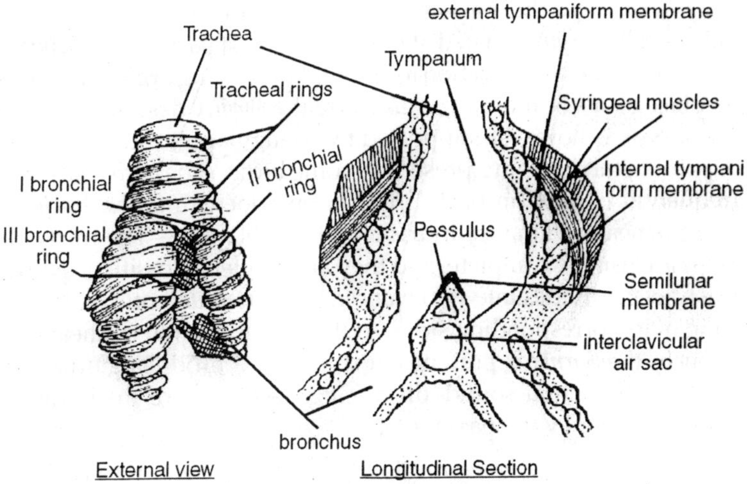

Fig.– Syrinx of Pigeon

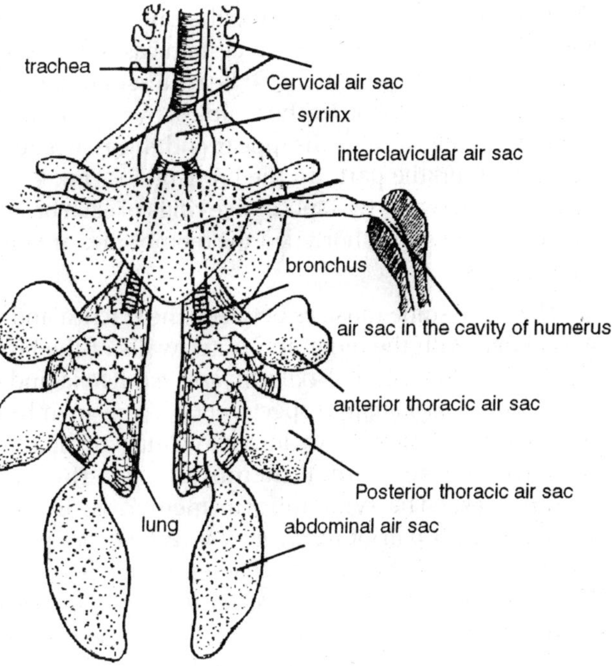

Fig.– Lungs and air sacs

Fig.30 : Respiratory organs of Fowl

is present near the base of the tongue and it opens into a rudimentary *larynx* situated at the anterior most part of the trachea. The larynx does not function as a sound producing organ in birds. Its place as a vocal organ is taken by the *Syrinx* described later. The larynx is however supported by triangular cricoid cartilage. A pair of *procricoids* are present in some birds. The *arytenoids* are frequently ossified in birds. The larynx opens into the *trachea.* The trachea is a long, cylindrical and flexible tube. The tracheal rings are usually complete, and they are frequently ossified. Now, the trachea divides into two *bronchi,* each *bronchus* enters into the lung of its corresponding side. At the division of the trachea into bronchi the *Syrnix* is present which is sound producing organ in birds. The actual sound–producing elements are membranes which vibrate by the passage of air.

The *broncho–tracheal syrnix* is very common. Here, the last rings of the trachea are united to form a resonating chamber known as tympanum. The folds of membrane, internal and external tympanic membranes extend into the cavity from the median and lateral wall of the bronchus. In certain cases, there is an internal skeletal element, the *Pessulus*. It bears a semilunar membrane on its lower surface. In the tracheal type of Syrinx the lateral portions of the last tracheal rings disappear and the associated membrane forms the vibratile part. In the bronchial syrinx the membranes are present between two successive rings of each bronchus. When the bronchial wall is shortened the membranes are forced as folds into the tube.

There are some muscles between the trachea and the bronchi. Associated with the inner and outer walls of the bronchus there are present the internal *tympaniform membrane* and the *external tympaniform membrane* respectively. These membranes are the mucous membranes. An interclavicular air space is enclosed beneath the semilunar membrane, by the tympaniform membranes. The tympaniform membrane is controlled and regulated by a pair of *intrinsic syringeal muscles* present between the trachea and the syrinx. These are inserted into the syrinx. There is one pair of *sternotracheal muscles* also arising from the sternum and inserting into the trachea. The sound is produced by the vibrations of the above membranes. The pitch of the voice is altered by the associated muscles i.e. *intrinsic syrengial* and *sternotracheal* muscles.

The Lungs

There is one pair of bright red and spongy lungs in birds. The bronchus enters into the lungs and is known as *pulmonary bronchus*. The lungs lie in the *pleural cavities*. The lungs in the case of birds are closely connected with the ribs and vertebral column and as such, they can undergo very little change in shape. The ventral surface of lung is covered by a peritoneal membrane known as *pleura*. The *costopulmonary muscles* are attached to the pleura. These muscles arise from the junction of vertebral and sternal ribs.

The bronchus enters the medioventral side of the corresponding lungs and loses its cartilaginous rings. The bronchus continues as a main trunk to the distal end of the lung and it known as *mesobronchus*. From the mesobronchus arise the secondary bronchi which are called *entobronchi, ectobronchi, laterobronchi* and *dorsibronchi* according to their position. These secondary bronchi give off small tubes of uniform diameter known as *prarbronchi*. Thus a series of parabronchial tubes with looped air capillaries carry on respiration instead of alveoli. There are 8 entobronchi, 6-10 ectobronchi, nearly 25 dorsibronchi (in chick) and a very few laterobronchi.

Air Sacs

Presence of air sacs is an important feature of the respiratory system of birds. Four secondary bronchi do no stop at the lung wall of each side, rather, they pass through the walls of the lungs and are continued as thin–walled sacs. The air sacs are thin walled, membranous, non–muscular and nonvascular structures present in all the living species of birds. There are *recurrent bronchi* leading from the air sacs to the parabronchi and the air capillaries of the lungs. In Pigeon the following air sacs are present —

a) *Interclavicular*: There is a single interclavicular air sac lying anterior to the furcula. From each side of this air sac arise two tubular air sacs known as axillary air sacs. One of these tubular sacs, the posterior one enters the humerus bone through a pneumatic foramen.

b) *Cervical*: There is one pair of cervical air sacs. The cervical sac of either side lies at the base of the neck. They are dorsal to the interclavicular sac. They have some small branches leading to the neck region.

c) *Anterior thoracic*: There is one pair of anterior thoracic air sacs lying at the ventral side of the lung. They are in close contact with the ribs. An oblique septum covers its ventral wall.

d) *Posterior thoracic*: There is one pair of posterior thoracic sacs also which overlap the posterior end of the corresponding lung.

e) *Abdominal*: There is one pair of abdominal air sacs each one arising from the distal end of its lung. They lie along the dorsal wall of the abdomen. It is ventral to the kidneys and is in close contact of the small intestine.

Functions of Air Sacs

The following are the important functions of the air sacs —

1. The air sacs act as balloons and reduce the specific gravity of the bird in air. They help in maintaining equilibrium during flight.
2. A complete change of air in the lung is made possible with every breath with the help of air sacs. Thus, they increase the availability of oxygen. They force air into the lungs at each expiration to renew the air.
3. As they extend between the muscles, they naturally reduce the friction during flight.
4. They aid in the inspiration and expiration processes, and, permits the air to pass twice over the respiratory surfaces of the lungs. It has been suggested that interclavicular, cervical and anterior thoracic air sacs are *expiratory* and they are supposed to be more active during flight. The posterior thoracic and abdominal air sacs are *inspiratory* and they are supposed to be more active when the bird is not flying.
5. They are helpful in reducing the bodily temperature in correlation with absence of sweat glands. The water vapour diffuses from the blood into the cavities of the air sacs finally passing out through the lungs.
6. The air sacs probably allow movement to the heart, which is rigidly set in the thoracic region.

Mechanism of Respiration

When the bird is at rest, the respiration is brought about by *Costopulmonary* or *intercostal muscles*. This muscle is inspiratory.

By the contraction of muscles the ribs are raised and the sternum area is lowered. Naturally, the area of thoracic and abdominal cavities is increased. As a result, pressure on the lungs is reduced and air is drawn in into the lungs. The air passes into the air sacs also.

During inspiration the path of entering air is,

nostrils → glottis → trachea → bronchi → mesobronchi → secondary bronchi → parabronchi → air capillaries.

Exchange of gases take place in the air capillaries.

But, the process of expiration is brought about by the movements of thoracic and abdominal muscles, and as such, these muscles are expiratory. The areas of thoracic and abdominal cavities decrease and as a result pressure on lungs increases. The air goes out of the lungs and finally through the external nostrils. By the compression of air sacs, the air goes into the recurrent bronchi and from there, it goes into air capillaries. This is actually a secondary supply of fresh air to the air capillaries.

During flight, the mechanism is quite different. The area of the abdominal and thoracic cavities is increased or decreased by the movement of the pectoral muscles of flight. The pressure of viscera against the air sacs causes air circulate. The respiration takes place also by the movement of sternum towards the vertebral column and away from it. When the speed of the bird is faster, rate of respiration is rapid.

Palate in Birds and their Relationship

The roof of the mouth cavity is known as *Palate.* The floor of the palate is formed by several bones e.g. *Basisphenoid rostrum, Basisphenoid* (or *Parasphenoid*), *Pterygoid, Palatine, Vomers* or *Pre–vomers* and *Palatine process of maxilla.* T.H. Huxley (1867) classified the birds on the basis of size and contacts of palatal bones and established the relationships of various types of palate in birds. The types of palates have been described below :

1. *Dromaeognathous*

The following are the important features of this type of palate—

(a) The pre–vomers (formerly called vomers) are large and broad and they extend far back. They articulate with posterior ends of the palatines and the anterior ends of the pterygoids.

(b) The palatines and pterygoids are separated from the parasphenoid (or basisphenoid of some authors).

(c) Basisphenoid is connected anteriorly with maxillopalatine (palatine process of maxilla) but the fusion of these bones does not take place.

(d) There are large basipterygoid processes arising from the basisphenoid which articulate with the pterygoids near their posterior ends.

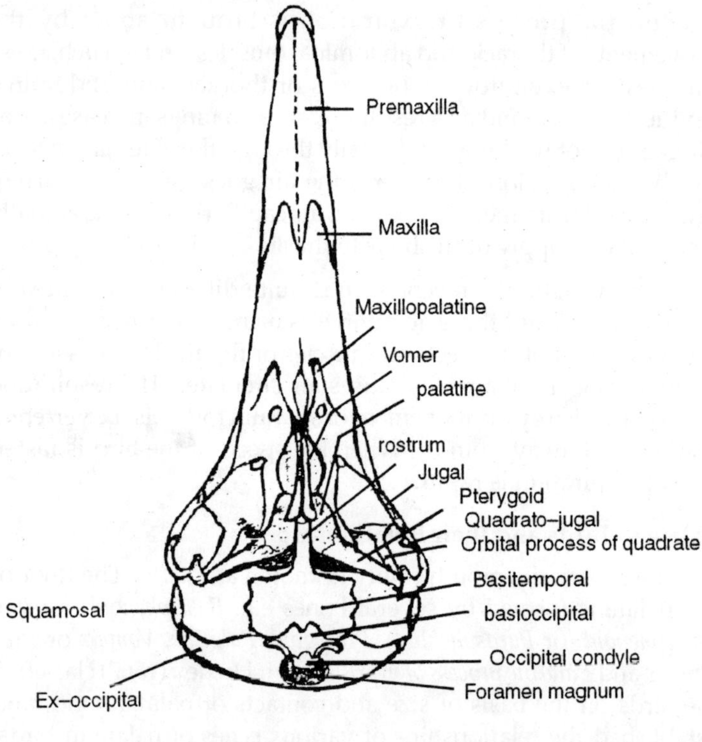

Fig.31: Dromaeognathous or Palaeognathous Palate

Huxley placed only *tinamous* (Tinamiformes) under this group but later on Pycraft (1900, 1901) showed that the palate of Rhea (Rheiformes) fits the same fundamental plan and should be included in the *dromaeognathous* group. Pycraft also brought all the struthious birds (Ostrich, Cassowaries, Apteryx etc.) under the group of "Palaeognathous" birds. But, Mc-Dowell (1948) could not be satisfied with this idea and this favoured four types of palates among these birds categorising them into *tinamiform, casuariform, struthiniform,* and *apterygiform.*

2. *Schizognathous*

   (a) Pre–vomers (vomers) if present are large or small and are completely fused. They lie at the tip of the rostum in fowl and are absent in pigeon.

   (b) Maxillopalatines do not meet along the central line, thus they remain separate.

   (c) Palatines and pterygoids articulate with the parasphenoid rostrum.

   (d) Pterygoids do not touch vomers in fowl.

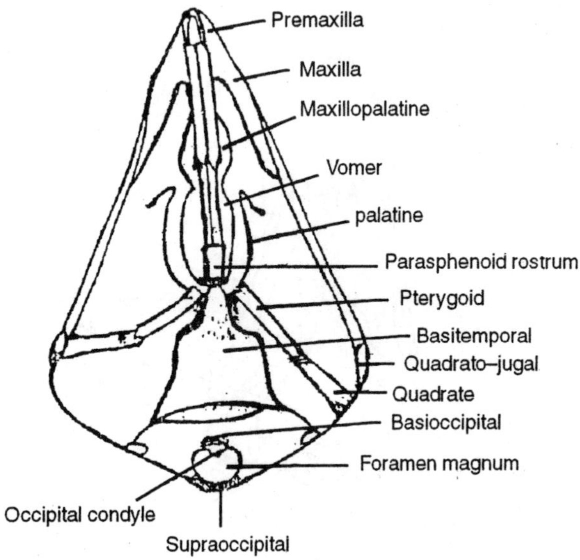

Fig.32 : Schizognathous Palate

This type of arrangement of palate is known as Schizognathous type and is found in turkeys (*Galliformes*), turnix (*Gruiformes*), sand pipers and plovers (both *charadriiformes*), wood peckers (*Piciformes*) and some other birds.

3. *Desmognathous*

(a) Prevomers (Vomers) are fused or absent

(b) They lie infront of the rostrum and are very small structures when present.

(c) Maxillopalatines meet in the mid–line infront of the vomers forming complete roof across the palate.

(d) In certain cases, palatines and pterygoids articulate with the parasphenoid rostrum.

(e) Pterygoids do not touch the prevomers (vomers).

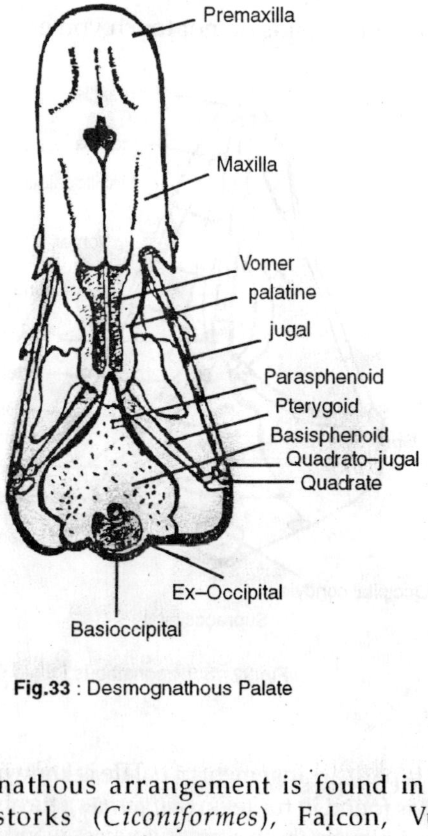

Fig.33 : Desmognathous Palate

This Desmognathous arrangement is found in Ducks (*Anseriformes*), storks (*Ciconiformes*), Falcon, Vultures (*Falconiformes*), Pelican (*Pelcaniformes*).

4. *Aegithognathous*

This type is similar to the *Schizognathous* type, but the main differences are —

(a) Prevomers are large, completely fused and truncated anteriorly. They are not pointed and slender.

(b) The maxillopaline is separated.

This type of arrangement is found in passerine birds (e.g. Sparrow).

W.K. Parker (1875) proposed a separate category to describe the woodpecker and this type was named as Saurognathous. This type was characterised by —

(a) Vomers paired, pointed and separate

(b) Pterygoids fail to touch the vomers

(c) Presence of a medio–palatine between the vomers.

That the vomers are separate has been criticized. The so called *Saurognathous* type has nothing to distinguish it from the *Schizognathous* type. The *Medio–Palatine* has been supposed to be fused vomers. Thus the proposed Saurognathous type is a superfluous one and it has not been maintained.

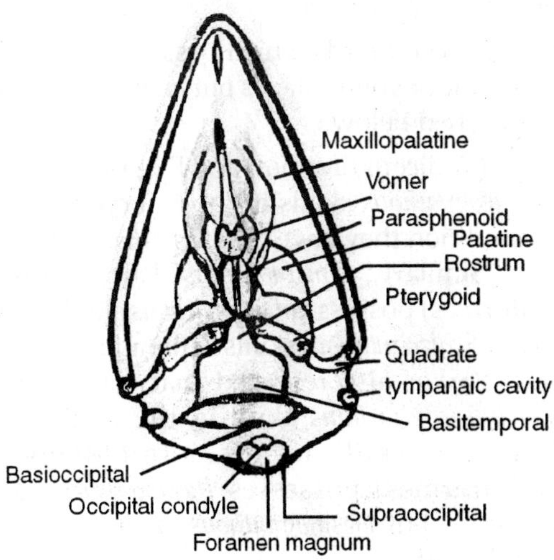

Fig.34 : Dromaeognathous or Palaeognathous Palate

Relationship of Various Types

It was recognised that the Dromaeognathous type is the only palate which is quite separate and distinguishable from other types. Rest palatal types like *Desmognathous, Schizognathous, Aegithognathous* and even *Saurognathous* merge into one another. For this reason Pycraft (1900) termed the *Dromaeognathous* type as *Palaeognathous* and the rest types as *Neognathous* type. Also, the *Aegithognathous* and *Schizognathous* type were considered as direct modifications of the *Palaeognathous* type, while the *Desmognathous* was derived from the *Schizognathous* type.

In *Palaeognathous* adults, the Pterygoid makes contact with vomers and has a long suture with the palatine but in the *Neognathous* adult, the pterygoid makes contact with the parasphenoid rostrum articulating by a joint with palatine. The pterygoid is excluded from the vomers. There are several *neognathous* birds which show *palaeognathous* features during embryonic development. In some neognathous birds the pterygoid bone in the embryo makes contact with the vomer, like adult Palaeognathous birds. But later on the front end of the pterygoid, becomes detached from the rest as *'hemipterygoid'* and fuses with the palatine. The separation of the pterygoid from the vomer by the palatine is the characteristics feature of the *neognathous* birds. The "hemipterygoid' bone of Pycraft was also called the "mesopterygoid" by Parker.

There are several birds which possessed *Palaeognathous* skull in the embryonic or young stages but *neognathous* skull in the adult stages as specified below :

Podiceps (Podicepidiformes) and Pygoscelies (*Sphenisciformes*) possess *Palaeognathous* skulls in the embryonic condition but in the adult condition they have *Schizognathous* skulls – as reported by Pycraft. Similarly, *Opisthocomus* (Falconiformes) and larus (Charadriiformes) possess *Palaeognathous* skulls in the embryonic condition but *Schizognathous* skulls in the adult stages – as reported by Parker. Parker also reported that corvus (Passeriformes) possesses *Palaeognathous* skull in the embryonic stage but *Aegithognathous* skull in the adult stage. *Steatornis* (Caprimulgiformes) possesses *Palaeognathous* skull in the embryonic stage but *Desmognathous* skull in the adult stage as reported by Pycraft.

Similarly, falco (Falconiformes) possesses *Palaeognathous* skull in the embryonic stage but *Desmognathous* skull in the adult stage as reported by Parker.

As we know, in *Palaeognathous* type the 'hemipterygoid' never becomes separated from the pterygoid. This type is retained in the adult form of certain birds e.g. Bucco (Piciformes). In certain cases, there is no hemipterygoid as in Gallus (Galliformes) and *Anas* (Anseriformes). The former has the *Schizognathous* skull and the latter possesses the *Desmognathous* skull.

As the *neognathous* birds pass through a *palaeognathous* stage in the embryonic condition, the "recapitulation theory" suggest that the *Palaeognathous* type is phylogenetically primitive.

But de beer (1937) is not in favour of the application of '*Recapitulation theory*' here. This author suggest that this dogmatic view should be discredited because the embryonic features may secondarily be retained in the adult stage. It has been pointed out that the *Palaeognathous skull* shows retarded development. Moreover, presence of basi–pterygoid process suggests a "kinetic skull" in which the pterygoid could easily slide.

Because of the presence of *neognathous palate* of *Schizognathous* type in *Archaeopteryx* it seems that the existing ratites are specialized nature rather than primitive. Variations are related to adaptive modifications. There are several evidences of embryonic characters being retained in the adult life (Neoteny) as may be deemed in the structure of the palate and other parts of the body. Palate does not provide a reliable basis for major classification.

Ratitae and Carinatae

Ratites form groups of palaeognathous (having primitive palate), flightless running birds. The ostrich and its allies are conveniently included in this group. The principal identifying characters of this group are that they are flightless running birds, with flat sternum and heavy legs quite suited for a terrestrial life. The ribs are without uncinate process and bone area without air cavities i.e. they are not pneumatic and are quite heavy. The wings are reduced or absent in these birds. The wing feathers comprise soft plumes or spike–like shafts which may be deemed to be adaptive features. These birds are well adapted to running in the deserts and savannas (extensive open grassy plains) of the old and new worlds.

Important characters of the various orders with examples have been mentioned below :

1. Order — *Struthiniformes* (e.g. *Ostrich, Struthio camelus*)

Characters

a) Ostrich is the largest living bird residing in African deserts and some other places. Males are about 8 feet high, having a weight upto 345 pounds.

b) They are flightless terrestrial birds which can run at a speed of 20–27 miles an hour.

c) They normally run with folded wings which are very small and have no interlocking arrangement of barbs and barbules. Feathers are quite soft.

d) Two wing feathers end in claws or spurs and are useful during attacks.

e) Rectrices are absent, only tail coverts are present. Feathers lack aftershaft.

f) There is reduction in the number of toes. There are only two toes (3rd and 4th) present on the foot.

g) The neck is long, much of which bare of feathers.

h) The head is flat and terminates into rounded beak with a huge gape.

i) The upper lid has prominent lashes.

j) Though they are mainly vegetarian but they are omnivorous.

k) Ostrich breeds in the spring season. The eggs are about 6 inches long and 5 inches in diameter, each weighing about 3 pounds. Incubation period is 42-48 days. The young one is precocial, about the size of the domestic hen. 12-15 eggs are laid per clutch.

l) Sternum is without keel and pubic symphysis is present.

m) There is no pygostyle.

n) The urinary bladder is present as it takes much water.

The ostriches which are found in deserts of Africa and Arabia were considered as the animal of desert like the camel and as such, it took its specific name *Struthio camelus* (ostrich camel).

2. Order – *Rheiformes* (e.g. Rhea, *Rhea americana*)

Characters

a) These are also flightless, swift running, terrestrial birds.

Fig.35 : Some flightless birds

b) It lives in flocks and feeds on vegetable matters and also insects and sometimes joins the herd of cattle.

c) The maximum height of the male if 5½ feet.

d) The plumage on the head and on the top of the neck are black in both the sexes. These are dirty white on the belly and yellow and gray on the rest part of the body.

e) The neck is partly feathered.

f) There are three clawed toes at each foot and they are webbed at the bases. Limbs are powerful.

g) Wings are more developed than the ostrich.

h) Sternum is without keel.

i) Upto 50 eggs may be laid by the female each egg weighing 2 pounds. The incubation period is 40 days.

j) Clavicle is absent.

3. Order – *Casuariformes* e.g. Emu (*Dromaius*) and Cassowary (*Casuarius casuarius*)

Characters

a) Emu (Dromaiidae) is the second largest living bird, male being 6 feet high. They are flightless birds.

b) The neck is partly feathered and it has three toes like *Rhea*.

c) Both male and female have deep brown plumage.

d) The feathers have two shafts at the base.

e) Wings are small or rudimentary.

f) Sternum is without keel.

g) 15 or more eggs may be laid by the female (in case of Emu) each weighing about 9½ ounces. The period of incubation is 2 months.

a) In Cassowary (*Casuarius casuarius*), the plumage looks like fur. There is a helmet or bony crest on the head. They are solitary birds, whose normal habitat is the forest.

b) The quills (on the wing) are reduced to stiff rods, and are hanging over the flanks.

c) The head and neck are partly bare but coloured yellow, blue, green, red and violet. Rest of the body is either black or brown.

d) Individuals are 4½ feet high.

e) There are three toes on each foot.

Emu resides in the great desert plains of Australia but Cassowaries are found in northern Australia, New Guinea and Polynesia.

4. Order — *Apterygiformes* (e.g. Kiwi, *Apteryx australis* – Common Brown Kiwi)

Characters

a) It is a flightless terrestrial bird.

b) It is about the size of the hen.

c) It has four toes in the legs.

d) The bill is long, pointed and curved having nostrils near the tip.

e) The wings are rudimentary or degenerate, and hidden beneath its body plumage.

f) The feathers are hairlike in appearance and they are without after shafts.

g) Both male and female are iron brown in colour and they are without tail.

h) The neck and legs are short and eyes are small.

i) They are nocturnal and feed on earthworms, insects, larvae, leaves and soft fruits.

j) They can run and jump with very high speed. The female lays a single egg of very large size.

k) Limb bones are not pneumatic.

The Kiwis are found in New Zealand and they are supposed to be rare survivors now. The scientific name *Apteryx* means '*without wings*'. There are two more species of *Apteryx* and they are Greater spotted kiwi (*A. haasti*) and the Little spotted kiwi (*A. oweni*).

5. Order — *Tinamiformes* (e.g. Tinamous, *Eudromia elegans* – Crested tinamous)

Characters

a) They are small, terrestrial, and cursorial creatures with very poor power of flight.

b) The skull is of palaeognathous type.

c) The tail in certain species possesses down feathers.

d) Sternum has keel in this bird

e) They are purely vegetarian

f) The eggs are glossy.

Most of the tinamous inhabit the rain–forests of South America. There are two very common species of this bird. They are Variegated Tinamous (*Crypturellus variegatus*) and Crested Tinamous (*Eudromia elegans*).

In addition to the above mentioned orders, there are two extinct orders also. These are — *Aepyornithiformes* (e.g. *Aepyornis*) and *Dinornithiformes* (e.g. *Dinornis*). The former includes the extinct flight-less elephant birds with vestigial wings, 4 toes in the legs, keel-less sternum and larger eggs. The latter includes the extinct birds of very huge size which are characterised by the presence of short beaks, aftershafts, large legs with 4 toes, reduced, keel–less sternum and absence of wings and pectoral girdles.

Carinatae birds

Characters

a) Unlike ratitae these birds have keeled sternum or carina and they are modern flying birds.

b) The carina is a bony ridge which increases the surface area of the sternum providing greater space for the paired flight muscles.

c) The flight muscles are much more powerful.

d) There is reduction of the clavicles.

e) There is a wide angle between the scapula and the coracoid of the pectoral girdle. These two bones are immovably articulated to form the glenoid cavity to receive the head of the humerus.

f) The clavicles support the upper ribs and strengthen the glenoid fossa.

g) Wings and tail are provided with flight feathers i.e. *Remiges* and *Rectrices*.

h) There is interlocking arrangement of barbs and barbules.

i) Bones are *pneumatic* and there is maximum fusion of bones to provide compactness of structure, necessary for flight.

j) The skull is of *neognathous* type.

k) Uncinate process of the ribs are present for the attachment of muscles.

Chapter 7

Class — Mammalia

All mammals are included in Class — Mammalia. In mammals (Gr. mammae = mammary glands), the body is covered with hairs which are epidermal in origin. In some aquatic mammals, there may be only a very few hairs. They are warmblooded (ectothermal or homoiothermal) animals. Certain glands, for example — sebaceous glands, sweat glands, scent glands and milk glands are present. Mammary glands are present in all the mammals for feeding the young ones. There are two occipital condyles in the skull. There are normally seven cervical vertebrae. Two pairs of pentadactyle limbs are present which serve various purposes. The toes are provided with horny claws, nails, hoofs or fleshy pads. In the lower jaw, a single bone dentary is present which articulates with the squamosal. The external ear or pinna is present with very few exception which has *external auditory meatus*. Pinna may be reduced in aquatic forms for making the movement convenient in water. Three ear ossicles, the *malleus, incus* and *stapes* are present in the middle ear. The teeth are *heterodont, thecodont* and *diphyodont* in most of the cases. The heart is four chambered and only left aortic arch is present. They are viviparous animals with the exception of prototherians (e.g. *Echidna*). A larynx with vocal cord is present. A muscular diaphragm is present to separate the thoracic cavity from the abdominal cavity. Erythrocytes are without nucleus, the exception being the camel. The fertilization is internal and placenta is formed (in viviparous forms) between the uterus and the developing embryo. Four optic lobes (*Corpora quadrigemina*) are present in the brain. There are 12 pairs of cranial nerves. Penis is also present in males.

Prototheria

Prototheria is considered as a sub–class under the class — Mammalia. The order — *Monotremata* of this sub–class comprises

Fig.36 : Some mammals with different adaptations

the most primitive mammals closely related to reptiles. Duckbilled platypus (*Ornithorhynchus*) and Echidna (*Tachyglossus aculeatus*) are the examples of monotremata. These are egg–laying mammals.

Ornithorhynchus is found in the rivers of Tasmania and south–eastern Australia. It is commonly known as Water Mole there. It lives in the burrows whose entrance is underwater. Its fur is reddish brown in the dorsal side and greyish in the ventral side. Duckbill is well adapted to aquatic life. The newborn young ones are weak and naked and they feed on discharged milk into the fold of mother's skin. There is sexual dimorphism in this animal. The male can be distinguished from the female by its greater size and also by the presence of spurs on the hind legs. The first specimen of this animal was discovered in New South Wales in 1797. Fossils are not available.

Tachyglossus (Echidna) is also called the *Porcupine Anteater* or *Spiny Anteater*. It inhabits the sandy and rocky parts of Australia, Tasmania and New Guinea. It is also a burrower. It captures ants with the sticky vermiform tongue. There is no sexual dimorphism. There are egg laying mammals and the eggs laid, are placed in the marsupial pouch where the mammary secretion is discharged. The young ones feed on the secretion in the beginning. Though this animal is clawed for making burrous, but it can move very fast.

General Characters of Prototheria

1. They are egg–laying (oviparous) mammals but nourish their young after hatching, with lacteal secretion.
2. The digestive and reproductive system have a single opening (the world montreme means 'one hole').
3. A long or short beak is present and the teeth are absent or degenerate.
4. A pectoral girdle with coracoid bones present
5. The coracoids reinforce the clavicles.
6. The interclavicle is a single T–shaped bone which joins the clavicle infront of the chest.
7. Mammary glands are without nipples.
8. There is one pair of rudimentary marsupial bones arising from the anterior edge of the pelvis into the abdominal muscles.

9. A single egg is placed in the pouch which (pouch) develops during the breeding period in case of Echidna. In platypus the eggs are laid in the nest.
10. The milk teeth (molars) have numerous cusps reminiscent of Mesozoic fossil. These milk teeth when shed, are not replaced by permanent set of teeth.
11. There is absence of Pinna.
12. Vertebrae are without epiphysis.
13. The body temperature does not exceed 83^{0}F.
14. The ribs have only a single head, no tuberculum.
15. The cervical ribs are present.
16. In the male each hind foot has a horny and mobile spur which is pierced by narrow canal leading from a gland of poisonous nature.
17. Tympanic bulla and lacrimals are absent.
18. The brain lacs *corpus callosum*.
19. The cochlea in not coiled.
20. Hairs are present on the body.
21. There are four optic lobes (*corpora quadrigemina*) in the brain.
22. The left aortic arch persists.
23. Epipubic bone is present.
24. Cerebral hemispheres are convoluted.
25. Echidna pterygoid bone is present.

Thus, is Echidna and Platypus there are not only mammalian characters but some reptilian characters also. They are definitely mammals because they possess mammary glands, hairs, four–chambered heart and only left aortic arch and other mammalian characters, but at the same time, they lay eggs, they have echidna pterygoid bone is skull, presence of cloaca and reptile like urinogenital organs, presence of interclavicle and other reptilian features.

Thus, monotremes are the survivor of an ancient animal type. They have probably been derived from a different group among mammal–like reptiles. Monotremes are primitive mammals still

Ornithorhynchus

Tachyglossus aculeata (Spiny ant–eater)

Macropus (Kangaroo)

Fig.37 : Examples of Prototheria and Metatheria

mid–way on its transitional journey from reptilian to mammalian organisation. The creation of the sub–class *Prototheria* under the class — Mammalia seems to be justified.

Metatheria

Metatheria is considered as an infra class under the class–Mammalia. This group (infra class) includes the pouch — bearing mammals which are placed in the order — *Marsupialia*. The female has an abdominal pouch, or *marsupium* in which the mammae are lodged. The young ones are borne in a rudimentary condition and further development takes place in the marsupium. The new borne offspring makes its way into the marsupium, climbing through the mother's fur after the birth. Marsupials are still far from the level of the placental mammals. Present–day marsupials are restricted in distribution and are confined to Australia and South America. They are the most numerous mammals of Australia occupying every part of the continent adapting to different habitats. The important examples of marsupials are :

(i) *Didelphis virginiana* (family – Didelphidae) (Virginian opossum)

(ii) *Dasyurus quoll* (family – Dasyuridae) (Australian native cat).

(iii) *Myrmecobius rufus* (family – Myrmecobiidae) (Australian banded anteater).

(iv) *Perameles* (bandicoot) (family – Peramelidae)

(v) *Notoryctes typhlops* (family – Notoryctidae) (Marsupial Mole)

(vi) *Trichosurus vulpecula* (family – Phalangeridae) (Brushtail opossum)

(vii) *Phascolarctos cinereus* (family – Phascolarctidae) (Koala)

(viii) *Vombatus hirsutus* (family – Vombatidae) (Common Wombat)

(ix) *Macropus rufus* (family – Macropodidae) (Red Kangaroo)

Important Characters of Metatheria

1. It comprises pouch bearing mammals.
2. True mammary glands are present which end in mammillae or teats.

3. The clavicles are not reinforced by coracoid bones.
4. The dentition is permanent, the monophydont condition present.
5. The anus is separated from the urinogenital apertures by a mascular wall known as *perineum*.
6. There are two vaginae opening separately into the urinogenital canal.
7. The oviducts are joined towards the end to form a vagina, the latter has a median partition.
8. The marsupials retain their eggs in the uterus until they hatch (viviparous).
9. There is no true placenta connecting the embryo to the mother.
10. The larynx of the young ones is extended by a tube to the nasal fossae for the purpose of free breathing during the semi–permanent suckling.
11. The epipubic (marsupial) bones are said to support the marsupium, but there is probability that they support the weight of the intestine.
12. The ribs bear two heads, tuberculum and capitulum.
13. Vertebrae possess epiphysis.
14. *Corpus callosum* is either poorly developed or absent.
15. There are two uteri (didelphic condition).
16. The jugal bones extend posteriorly and take part in the formation of the glenoid cavity.
17. Dentary has an inflected angle.

Concluding Remarks

In the light of the characters discussed above, the marsupials are definitely more advanced than the monotremes. They have been removed from the common stem of mammals, birds and reptiles. Marsupials were the dominant mammals in the mesozoic era. Marsupials have been divided according to their dentition into two sub–orders, the *Polyprotodontia* and the *Diprotodontia*. The Polyprotodontia includes the opposums, dasyures, thylacines and bandicoots. Similarly, the Diprotodontia includes Kangaroos, Phalangers, Koalas and Wombats. In the former group several canines may be present, but in another group the canines are rudimentary or absent and molars have fine tubercles. Marsupials decidedly deserve a higher zoological position than the monotremes. Their small size and arboreal habit enabled them to keep out of reach of the great reptiles. Later on, they disappeared in many parts of the world probably in competition with the more advanced placental mammals. Their inclusion in sub–class theria and infraclass – Metatheria seems to be justified.

The Primates

Primates is an important order of class – Mammalia and man belongs to this order. There is general agreement that the primates arose from very early arboreal insectivores. They are generally omnivorous and arboreal mammals with completely hairy body. Hands and feet are nearly prehensile. Limbs are pentadactyle and they are distinguished by the flattening of claws into nails. Thumb and toes are shorter and opposable to other digits suited for grasping. The stomach is simple but sacculated in certain forms. The locomotion is plantigrade. Orbits are directed forward and surrounded by bony rings. Tympanic bulla is present. The brain is voluminous and complicated with well developed convoluted cerebrum. Clavicles are present. There is normally lac of 3rd trochanter in the femur. Testes are present in the scrotum and the penis is pendent. There are normally two pectoral mammae with teats. The placenta is discoidal or metadiscoidal sometimes diffuse, and haemochorial.

Classification with Important Characters

The order – Primates has been divided into three sub–orders i.e. *Lemuroidea, Tarsioidea* and *Anthropoidea*. They have been described below with characters and suitable examples :

Sub–Order — *Lemuroidea*

General characters : They are fox like in appearance with low cranium. The hand and tail are not prehensile. Though the second digit of the foot is clawed but the rest of the fingers have flat nails. The fore limbs are shorter than hind limbs. Pectoral mammae are present. They are arboreal and mostly nocturnal. The head has sharp muzzle. The temporal fossa opens widely to the orbit. A third trochanter may be present in the femur. The cerebral hemispheres are relatively small. The turbinal bone is well developed. The uterus is bicornuate and the placenta is diffuse.

The lacrymal foramen lies infront of the orbit and jugal extends backwards. Lemuroidea is divided into two families —

Family 1. *Lemuridae*: The animals are characterised by the presence of a free tympanic annular ring in the bulla. They have fox like muzzles covered with thick hairs. The brain is not well developed. The dentition is adapted to fruit diet. Lemuridae comprises four sub–families —

A. Sub–family — *Lemurinae* — It includes true *Lemur*. It has thick wooly fur of different colour combination. The skull is long. The dental formula is –

$$\frac{2, 1, 3, 3}{2, 1, 3, 3} = 36$$

B. Sub–family — *Indricinae* — This sub–family includes *Indris, Propithecus* and other forms. Ears are hidden beneath the silky fur. These are diurnal animals. The skull is short and there are 30 teeth which is a characteristic feature. They are herbivorous and the molars have tubercles. The dental formula is –

$$\frac{2, 1, 2, 3}{1, 1, 2, 3} = 30$$

C. Sub–family — *Cheirogalinae* — This sub–family includes *Cheirogaleus, Microcebus* and other forms. *Cheirogaleus* is about the size of a rat and *Microcebus* is known as Dwarf Lemur because of its smaller size. There is one pair of fenestrae in the palate. The skull is short. The tarsals are quite long.

D. Sub–family — *Cheiromyinae* — Aye–aye (*Daubentonia madagascariensis*) is the only genus of this sub–family. It is squirrel like in appearance, about the size of a cat. It has thick silvery fur and a long and bushy tail. The skull is short. They are nocturnal tree climbers. The third finger is very long, no thicker than a wire with which it finds insects in the barks.

Family 2. *Lorisidae* : This family includes the lorises. Lorises are the inhabitants of India, Ceylon, South–east Asia and East Africa. There are two sub–families under this family which have been briefly described below :

A. Sub–family — *Loricinae* — *Loris, Arctocebus* and *Nycticebus* are the important examples of this sub–family. They are short sized animals and they have no tail. The index fingers are vestigial. Their movement is slow and they creep about the trees at night.

B. Sub–family — *Galaginae* — *Galago crassicaudatus* of East Africa is a very common example of this sub–family. It is over one foot and has a longer tail. The muzzle is short and the tarsal is elongated. Two thoracic and two abdominal mammary glands are present.

Sub–Order — *Tarsioidea*

This sub–order includes the tarsiers which are found in Borneo, Sumatra and Philippines. *Tarsius spectrum* is a very common example. It is arboreal, nocturnal and insectivorous. The external ears are very large and mobile. The eyes are large and protruded and dominate the face. The face is flat and the skull is round. Orbits of the eyes are comparatively separated from the temporal fossa. It is about six inches long and tail is about 10 inches in length. Hairs are present only on the terminal part of the tail. Tarsal bones are elongated. *Tarsius* is quite specialised in jumping. The tympanic ring remains outside the bulla and forms the external auditory meatus. Jugals do not extend far back. The brain is relatively large but the olfactory part is comparatively small.

The uterus is bicornuate and the placenta is discoidal with haemochorial organization. Some digits of hind limb (2nd and 3rd) have retained the ancestral claws. Fingers and toes form the adhesive pads. The dental formula is –

$$\frac{2, 1, 3, 3}{1, 1, 3, 3} = 34$$

Sub–order — *Anthropoidea*

This sub–order includes primates with flattened bare faces. The eyes are small with orbits completely closed laterally. The cranial cavity is large. They are terrestrial, arboreal and diurnal. Teats are restricted to thoracic area. Teeth are 32 to 36 in number. Cerebral hemispheres are well developed covering the cerebellum and medulla oblongata. Fingers and toes are provided with flat nails. Anthropoidea has been conveniently divided into two groups as described below with the characteristic features.

Group 1. *Platyrrhini* —

It includes the new world monkeys. This group is characterised by the presence of broad nasal septum, 36 teeth and prehensile tail. They are less intelligent. The tympanic bone is a ring like structure, tympanic bulla is present but external auditory meatus is absent. The dental formula is –

$$\frac{2, 1, 3, 3}{2, 1, 3, 3} = 36$$

Cheek pouch is not present. The nostrils are directed outwards.

Group 2. *Catarrhini* —

It includes the old world monkeys. The internasal septum is narrow and the nostrils are directed downwards. The number of teeth is 32 and the dental formula is –

$$\frac{2, 1, 2, 3}{2, 1, 2, 3} = 32$$

The tail, if present, it is always penhensile. Cheek pouch is present except in apes and some other members. Members of catarrhini are more intelligent than that of the Platyrrhini.

Now, the group *Platyrrhini* (New world monkey) has been divided into *Haplidae* and *Cebidae* which deserve the status of super–families. Marmosets (e.g. *Callithrix*) have been placed in *Haplidae*. Marmosets show some primitive characters like – claws on every digit except the hallux. They possess bushy, squirrel like tail. There has been atrophy of wisdom teeth and thus, the number of teeth has become 32 only $\left(\frac{2, 1, 3, 2}{2, 1, 3, 2}\right)$. The tail is non-prehensile having soft fur. The pinnae are hairy. The pollex is not opposable.

On the other hand, in *Cebidae,* all digits end in nails and there is no trace of claws. Pinna is not hairy. The tail is long and prehensile. The dental formula is – $\left(\frac{2, 1, 3, 3}{2, 1, 3, 3}\right) = 36$. The hallux is opposable. Cebidae comprises four families which have been described below :

Family 1. *Mycetinae* —

It includes the howling monkeys (e.g. *Alouatta caraya*) which are the largest of the new world monkeys. It has long and prehensile tail. The lower incisor is vertical and pollex is well developed. The larynx is well developed and howling voice can be heard from miles away.

Family 2. *Pithecinae* —

Pithecia sp. of the Brazilian forests possess curly hairs and bushy tail. The tail is long and prehensile. Pollex is well developed. The lower incisor is inclined forward.

Family 3. *Nyctipithecinae* —

It includes the squirrel monkeys (e.g. *Saimiri sciureus*). The skull is comparatively large and the brain is quite big. The tail is long, either non–prehensile or slightly prehensile. Pollex is well developed and the lower incisors are vertical.

Family 4. *Debinae* —

This family includes the spider monkeys (e.g. *Ateles ater*). Most of the members of this family can live well in captivity. They have flesh coloured face. They easily learn to imitate man's action. The tail is long and prehensile. Pollex may be absent in some members. The lower incisor is vertical.

Catarrhini (old world monkeys) can also be conveniently divided in to three families which have been described below :

Family 1. *Cercopithecidae* —

In this family the members possess close nostrils and the internasal septum is very thin. There is well developed thumb. Arms are shorter than the legs. Cheek pouches are present, in which the food can be stored. The stomach is divided and the caecum is conical without vermiform appendix. The sternum is long and narrow. This family has two sub–families.

A. Sub–family — *Cercopithecinae* — The Northern Indian and South–east Asian species, *Macaca mulatta* is a common example of this sub–family. It is intelligent and curious. They are gregarious and agile. They soon learn to manipulate simple tools. The tail is long and reduced and the stomach is globular.

B. Sub–family — *Semnopithecinae* — Langurs (e.g. Presbytis) and colobus monkeys (e.g. *Colobus abyssinicus*) are included in this sub–family. The body is generally slender, there is reduction in the thumb of *Colobus*. Long silky fur can be seen hanging their sides like ornaments (in *Colobus*). The tail ends in beautiful long tuft. The stomach is sacculated and there is no cheek pouch.

Family 2. *Pongidae* — Four important examples have been described below :

A. Gibbons (e.g. *Hylobates lar*) : They can walk upright on the ground. They have very long arms. Fingers are hooked inwards (i.e. towards the palm). They are slender and swift moving animals. They can produce long voice. They are arboreal.

B. Gorilla (*Gorilla gorilla*) : They are very strongly built and they have tremendous strength. Ears are small and they eyes are deeply sunk. The body is so heavy that it can not lead an arboreal life. The face is naked and black. Limbs are more like human in proportions.

C. Chimpanzee (e.g. *Pan satyrus*) : They are less arboreal than gibbons. Ears are large. They tend towards man in its brain, dentition and size of the limbs. Arms and legs are almost of the same size. Similar blood reactions and acceptance of grafted tissues show that they are close to man. They are less bulky.

D. Pongo or Orang–uṭan (e.g. *Pongo pygmaeus*) : The head has developed upwards producing a high arches fore head. Arms are longer than gorillas and chimpanzee. They live in swampy forests. The size of the canine is larger.

Family 3. *Hominidae* —

Homo sapiens, or modern man is a member of this family. Man is characterised by —

(A) Large brain and high intelligence (cranial capacity 135 – 2000 C.C.)

(B) The forehead is high

(C) Forelimbs evolved into efficient manipulating device for making tools and to show all sorts of creativeness

(D) Bipedal locomotion

(E) Articulated speech

(F) Upright posture

(G) Eye ridges absent

Hylobates (Gibbon)

Pongo satyrus (Orang–utan)

Macaca mulatta

Fig.38 : Some arboreal mammals

(H) Orthognathous face with a chin etc.

(I) Lac of prehensile tail

Human population has been divided into races e.g. *Negroid race, Mangolian race,* and *White race*. The Negroid race has dark (black) skin, broad flat nose, curly hairs, thick lips and narrow pelvis. Similarly the Mangolian race is characterised by yellow skin, black and straight hairs, small nose, broad face, swollen eyes. The white race has narrow but prominent nose, small teeth, broad pelvis, soft and straight hairs, and retreating cheek bones. The above mentioned races have been further divided into varieties.

Distribution of Some Important Primates

The evolution of primates took place both during the *Mesozoic* and *Cenozoic* eras. Lemurs are very common inhabitants of Madagascar. Lorises are found in India, Ceylon and South–East Asia. The Great Galago is found in East Africa. Some species of *Golago* (e.g. *G. senegalensis*) are found in Ethiopia and Mozambique. *Indris* has been reported from Madagascar. Tarsier inhabits in the forests of Barneo, Philippines and Sumatra. Titis (e.g. *Callicebus torquatus*) are found in Brazil, Peru and upper Amazon. Several howling monkeys inhabit tropical America. Marmosets are found in Brazil. Rhesus Monkey widely spread in northern India and South–east Asia. Pig–tailed Macaque has been reported from Assam, Burma, Malaya, Thailand and East Indies. The Asian genus *Macaca* is found in China, Japan, India, Burma, Malaya and also in North Africa and Gibraltar. Baboons inhabit south, west and central Africa. The Gibbons are found in Burma, Thailand, Indo–China, Java, Borneo and Sumatra. The Gorilla is found in the forests of Cameroun, Gobon and the Congo and in the mountains north of Lake Kiru, and Uganda border. Urang–utans (Pongo) inhabit the forests of Borneo and Sumatra. The Mandrill (*Mandrillus sphinx*) and Chimpanzees are confined to Africa.

Cetacea

Cetacea is an order under the class – Mammalia. Cetaceans are the mammals completely adapted to an aquatic life. Comparative anatomy shows that they have descended from terrestrial animals. But, their ancestry is still very difficult to understand. The earliest cetaceans are known from the Eocene but the modern cetaceans appeared suddenly in the early part of the Miocene.

General Characters of Cetacea

1. They swim well and they are streamlined.
2. In dermis there is a uniform layer of fat known as *blubber* which conserves heat and is a source of energy.
3. There is no external ears, scrotum and projecting nostril which would have hampered movement through water. The opening leading to the ear–drum is filled with a *wax plug*.
4. There is scarcity of hairs on the body, only a few hairs may be present on the muzzle and in some whales they may be modified into whale bones.
5. The neck is very short, head being continuous with the body. The fore–limbs or flippers are well developed. The flippers have the same basic skeleton as found in the fore-limbs of terrestrial mammal.
6. The tail is the real locomotory organ. It has two horizontal flukes projecting on either side. The tail acts as a propeller and as a horizontal rudder regulating vertical movement in water.
7. The hind limbs are absent, however, vestiges of a pelvis and sometimes of a femur and tibia may be present inside the body.
8. The stomach has a masticatory and a secretory part.
9. The lungs are of enormous size which can hold 100 to 140 cubic feet of air i.e. about 28 cubic feet of pure oxygen which can last for a longer period. There is also much larger number of alveoli. The bronchial passage is provided with sphincter. The lungs are non–lobulated and the liver is bilobed. Fatty foam is present in the respiratory tract which is excretory. There is well developed system of *retia mirabilia*.
10. The larynx opens directly into the nasal fossae.
11. The right side of the heart beats slowly and sends little deoxygenated blood back to the lungs but the left side maintains long supply of oxygenated blood to vital organs like cerebrum, cerebellum and medulla oblongata.
12. Great obliquity and powerful musculature of diaphragm is quit remarkable.

13. The brain is small and the nasal passage has almost lost the sense of smell. The auditory passages are narrow. The small eyes are well adapted to the refractive index of water. Eye lids are rudimentary and the secretion of the lacrymal glands protects the cornea and conjunctiva from the action of sea water.
14. The testes are internal and the penis is placed in a groove. The uterus (in female) is bicorned and diffuse in its nature.
15. In Mysticeti the mouth accommodates the plates of whale-bone or baleen which take the place of teeth.
16. The tongue is muscular.

Skeletal details of Cetacea

Vertebrae :– 1) The cervical region is comparatively short. There are seven cervical vertebrae which are very much compressed and often fused into a single disc. Because of the formation of a continuous mass, the head can not be moved in relation to the body.

2) The neural spines of the vertebrae are well developed.

3) When the vertebrae are separate, they have small arches and long transverse process having two narrow bars.

4) The epiphyses are very distinct discs.

5) There is poor development of zygapophyses which may also be absent particularly in the posterior portion of the trunk.

6) The caudal region consists of numerous vertebrae and there is a series of chevron bones.

7) There is no sacral region in the vertebral column.

Ribs and Sternum

1) In the baleen whales (Masticeti), only one pair of ribs articulate with the sternum.

2) In the toothed whales (Odontoceti) distinct tubercles and heads are present. The tubercles articulate with the transverse process and the heads with the vertebral bodies. There is no heads in the hinder ribs and as such only the tubercle articulates with the transverse process. The anterior ribs are attached with the sternum by sternal ribs which are perfectly ossified.

3) A presternum is also present sometimes. The Xiphisternum is absent.

Pectoral Girdles and Fore-limbs

1) The scapula is simplified. It is broad and flat like a fan and is without a spine.
2) The acromian is slightly curved laminar process with a narrow base.
3) The coracoid is compressed and parallel with the acromian but, in certain cases both coracoid and acromian may be absent.
4) The clavicle is absent.
5) The humerus, radius and ulna are relatively reduced in size. The hand is elongated by the multiplication of phalanges. The digits are enclosed in fibrous tissue. Because of *hyperdactyly* and *hyperphalyngy* the surface of the flipper increases.
6) In the toothed whales the carpels are fully ossified.

Pelvic Girdles and Hind-limbs

The pelvic girdles are present in the form of vestiges. Parallel to the spinal column, a pair of long narrow bones are present. They represent the ischia. Vestiges of tibia and femur are also present.

Skull

1) The skull is rounded and the jaws are greatly elongated.
2) Supraorbital plate is quite large which develops from the frontal bone.
3) Usually the maxilla is very large extending backwards in the toothed whales and it overlaps the frontal. In the baleen whales the extension is below the orbital process of the frontal bone.
4) The jugals are very small bones but the zygomatic process of the squamosal is large and stout.
5) The premaxillae are long and narrow bones, the nasal is small. The tympanic bone is quite large. In Baleen whales two bony pedicels connect it with the periotic. The pedicel is absent in the toothed whales.

Teeth

Teeth are present in the sub–order — Odontoceti. In dolphins and porpoises 100 to 200 homodont conical teeth are present. In the sperm whale, the functional teeth are present only in the lower jaw. Very few teeth are present in the beaked whales. It is remarkable that there may be only one tooth in each half jaw. The narwhal has only single tooth developed as a tusk in the upper jaw.

Brief Classification of Cetacea

Now it is a well established fact that — Cetacea is an order under the class — Mammalia. This order comprises two sub–orders i.e. *Mysticeti* and *Odontoceti*. Mysticeti comprises three families and Odontoceti has eight families. They have been described below with important characters and examples :

Sub–order — *Mysticeti*

Family 1. *Balaenidae* —

Characters —

(1) There is absence of dorsal fin and grooves on throat and chest of the animals belonging to this family.

(2) The two halves of the upper jaw are strongly arched. They form a 'V'–shaped structure. They hold about 300 baleen plates on each side.

(3) The lower jaw is also V–shaped and horizontal.

(4) The lower lip rises to meet the upper lip and covers the whalebone.

This family includes the right whales which comprises three genera — *Balaena, Eubalaena* and *Neobalaena*. They are commonly known as whalebone whales.

Family 2. *Balaenopteridae* —

Characters —

(1) In this family the dorsal fin and ventral grooves are necessarily present.

(2) The head is small and the jaws are not arched.

(3) As the baleen plates are shorter, the jaws are less wide apart.

(4) Unlike balaenidae the lower lip does not come up so high. Also, the flippers are much more pointed.

This family comprises the Blue whales, *Balaenoptera musculus*, fin whale *B. physalus*, and Sei whale *B. borealis* etc. The blue whales are the largest animals on the earth which may reach a length of up to 100 feet with a weight equivalent to that of twenty four elephants. They are commonly known as Rorquals.

Family 3. *Rhachianopteridae* —

Characters —

(1) There is no dorsal fin but two ventral grooves are present.

(2) Its length is about 45 feet.

The Californian Grey Whale, *Rhachianectes glaucus* is a common example of this family.

Sub–order — *Odontoceti*

Family 1. *Physeteridae* —

Characters —

(1) These are toothed whales. Teeth are present in the lower jaw only. The total number of teeth is only 40, each one about 8 inches long.

(2) They have massive blunt snout in the cavity of which, a thin, colourless transparent oil is present. This oil solidifies when brought in contact with air. This solid white wax is called *Spermaceti*. In the intestinal concreation an another substance is found which is known as *ambergris*. It is used in the preparation of perfumes.

(3) The real dorsal fin is lacking but a series of low humps in the hindmost part of the body.

(4) They are black above, grey on the sides, and white below.

They are known as sperm whales and the common example is *Physeter catodon*.

Family 2. *Kogiidae* —

This family has been created to accommodate the Pigmy Sperm whales of Atlantic, Pacific and Indian Oceans. Most of the characters are like that of the Sperms whales.

Family 3. *Phocaenidae* —

Character —

(1) It has a low dorsal fin.

(2) Though the flippers are small, however it is an efficient swimmer.

(3) It possesses 100 teeth.

(4) They feed on fishes sometimes.

(5) Mating takes place in the summer and gestation period is one year.

They are commonly known as porpoise. The smallest cetacean of the European waters, *Phocaena phocaena* is the best example of this family.

Family 4. *Delphinidae* —

This family includes the Grampuses and dolphins.

Characters —

(1) The Grampus or Killer Whale, *Orcinus orca* is the largest cetacean of this family. The male is about 30 feet long but the female is just half in length.

(2) The dorsal fin of the Killer Whale is very high and the flippers are broad and rounded.

(3) Though, the whole body is black but there is a white patch over the eye.

(4) *Grampus griseus* is also a good example which is found off the coast of Europe. The forehead is rounded and the flippers are long and pointed.

(5) The common dolphins, *Delphinus delphis* has a prominent beak separated from the snout by a groove.

Family 5. *Ziphiidae* —

Characters —

(1) This family includes the Bottle nosed Whale (*Hyperoodon rostratus*) and the Beaked Whale (*Ziphius cavirostris*) as its best examples. Both species are cosmopolitan.

(2) In the beaked whales the teeth are present in the lower jaw. The beak is the elongated snout.

(3) The teeth are vestigial except one or two pairs. The teeth are embedded in the gums in the young ones of both sexes and in the females of all ages.

(4) The length varies from 13–40 feet in different species.

Family 6. *Monodontidae* —

Characters —

(1) *Monodon monoceros* commonly called the Narwhal, is the only species in this family.

(2) They normally inhabit arctic waters.

(3) The skin is white but there are black spots on the body.

(4) The dentition is singular.

(5) All the tooth buds are present in the embryonic stage but they disappear in the adult. Only two incisors are present. In females these are undeveloped but in the males, one of the incisors grows enormously, forming a twisted tusk reaching a length of as long as upto eight feet. Development of tusk in females is not a normal case.

(6) The Narwhal is not more than 15 feet in length excluding the tusks.

Family 7. *Platanistidae* —

Characters —

(1) These are river dolphins, thus adapted to freshwater habitat.

(2) They vary in length from 5 to 10 feet only.

(3) The long bird like beak is furnished with teeth.

(4) The Gangetic Dolphin, *Platanista gangetica* is found in Brahmaputra, Indus and Ganges. It is blind, the eyes being mere vestiges.

Family 8. *Iniidae* —

This family includes the Amazonian Dolphin *Inia geoffrensis* and the Chinese River Dolphin, *Lipotes vexillifer*. They were discovered later on.

Chiroptera

Chiroptera is an order under the class — Mammalia. This order includes bats and flying foxes capable of active flight. Fossils occur in lower Eocene and they were like modern forms. They have been derived possibly from an insectivore – like ancestor. Forelegs have modified into wings so that they can lead a volant life. Each wing is a thin sheet of skin between the fingers (except the thumb) reaching the side of the body, hind legs and tail. Wings of courser resemble *Pterodactylus*. The order chiroptera contains about 750 species.

General Characters —

1. The forelimbs are modified into wings which are richly supplied with blood vessels.
2. The fore–limbs are elongated. The fold of the skin is supported by 4 ulnare digits. This fold is known as *Patagium* which begins from the shoulder and extends along the upper margin of the arm to the base of the greatly elongated thumb. Then it passes between the long fingers and along the sides of the body to the hind limbs. The *interfemoral membrane* between the hindlimbs may or may not involve the tail.
3. The knee is directed backwards and the *hallux* became external and this condition is achieved because the hind limb has rotated.
4. Some bats have elongated muzzle and the pinna becomes slightly complicated by the presence of an inner lobe, *tragus*.
5. Elongated leaf like lobes are present in the nose.
6. In majority of the bats the body is covered with fur, however, some bats are completely naked particularly in the sub–order microchiroptera.
7. The vertebral column is comparatively short and there is mobility between the vertebrae.
8. The neural spine is absent in cervical and thoracic vertebrae.
9. The transverse process in the whole lumbar region is rudimentary.
10. Normally, there are 12 thoracic and 5 lumbar vertebrae.
11. Ribs are flat and the thoracic frame work is rigid.
12. Development of sagittal and occipital crest in the skull can be well marked.

13. The facial region is elongated.
14. The post orbital process is either very short or absent.
15. Nasals are long and narrow bone of the skull.
16. The Zygomatic arch is long and slender.
17. Angular process is absent in most of the bats.
18. Premaxilla if present, is cartilaginous.
19. The sternum has a keel for the attachment of flight muscles.
20. They are nocturnal and gregarious.
21. The pectoral girdle is strong with large and oval scapula and well developed acromian process.
22. The clavicle is long and curved and a coracoid process is present.
23. The humerus has long pectoral ridge and there is no supra–condylar process.
24. The ulna bone is rudimentary and is fused with the radius.
25. The pelvic girdle is small with imperfect symphysis.
26. Long and strong sacrum unites with the ischia.
27. The fibula may be either well developed or reduced.
28. A process, known as *tubercalcenei* develops from calcaneum to which the calcar is attached with the help of ligaments.
29. The dental formula in a horse–shoe bat is —

$$i\frac{1}{2},\ c\frac{1}{1},\ p\frac{2}{3},\ m\frac{3}{3} = 32$$

30. The alimentary canal is marked by a narrow oesophagus, short intestine. A short caecum is present in the frugivorous bats, it is well developed in blood sucking bats and absent in some other bats.
31. In *Desmodus* there is a pyloric diverticulum which is usually filled with coagulated blood sucked from cattle.
32. Cerebral hemispheres do not cover the cerebellum and the spinal cord is narrow behind the neck.
33. Sense of touch is also remarkably present in the wing skin.

34. The large pinnae, distal hairs of snout and the nose leaves are sensitive and the tragus is present which is associated with the ear.

35. They are noctural creatures with only rods in the retina.

36. The flight of bats is directed by a radar like system using ultrasonic waves outside the range of human ear. According to Galambos and Griffin the ultrasonic sounds are emitted at a rate of ten to the second when the animal is at rest and thirty to sixty at the time of flying. The wave frequency as they recorded was found to be between 50000 to 100000/ second.

37. Unlike other mammals, the larynx has powerful muscles inserted on bones.

38. Most of the bats sleep in clusters.

39. Some bats can migrate to avoid excessive cold e.g. *Lasiurus borealis* of North America.

40. The penis in males is long and pendent, the penial bone may be present in certain species.

41. Mating takes place once in a year and spermatozoa can be retained in the uterus for a longer period and fertilization is delayed. Normally, only one young one is given birth at a time. The young one is naked and blind. The milk teeth are replaced by permanent teeth after sex weeks. Gestation period is sixty days in bats.

Classification of Chiroptera

Order–Chiroptera has been divided into two sub–orders which have been described in detail with suitable examples below :

1. Sub–order — *Megachiroptera*

These are large fruit–eating bats. The molars are flattened. The second digit has 3 phalanges and it is clawed. The snout is elongated and there is no foliaceous appendages (tragi or nose leaves). The pyloric part of the stomach is much elongated. The mandible has no angular process. The tail is not enclosed in interfemoral membrane.

The common examples are —

The Indian Flying Fox (*Pteropus giganteus*), the South African Hammerhead Bat (*Hypsignathus monstrosus*), and the long-tongued fruit bats (*Macroglossus* and *Megaglossus*) of India, New Guinea, Africa and Australia. The Kalong (*Pteropus vampyrus*) of West Indies is the largest bat measuring 16 inches in length and having a wingspan of over 60 inches.

2. Sub–order — *Microchiroptera* —

These are small and medium sized insectivorous bats. The molars have cutting crests. The stomach is simple. The snout is short having foliaceous appendages. The tail is enclosed in the interfemoral membrane. The mandible is provided with angular process. The second digit has one or two phalanges and it is with claw. This sub–order has certain important families described below :

Family (1) *Megadermatidae* : This family includes the false vampires. They possess nose leaves and tragi but there is absence of tail. The common example is, *Megaderma gigas*.

Family (2) *Rhinolophidae* : This family includes the Horseshoe bats. They are characterised by the presence of nose leaves and a tail but they have no tragi. The common examples are — Greater Horseshoe Bat (*Rhinolophus ferrum–equinum*) and Lesser Horseshoe Bat (*R. hipposideros*). The word horseshoe indicates the shape of their nose leaves. The Horseshoe bats are found in different parts of Europe, North Africa and Asia. The Lesser Horseshoe bat is nearly half the size of the Greater Horseshoe Bat.

Family (3) *Vespertilionidae* : This family includes the long–eared bats and allies. The long–eared bat (*Plecotus auritus*), Mouse–eared Bat (*Myotis myotis*) and *Pipistrellus subflavus* are the common examples of this family. In *P. auritus*, the ears are as long as the body. In summer it lives alone or with its mate in the hollows of the trees. They can walk and climb. They are small bats found in Eastern Europe, the United States and Canada. The European *Myotis myotis* has a wingspan of about fifteen inches. The tragi are long and pointed. They are very fast flyers. *P. subflavus* is the smallest American bat which is not more than three inches long.

Family (4) *Phyllostomatidae* : This family includes the American leaf–nosed bats. Some of the important examples of this family are — *Macrotus californicus*, *Vampyrus spectrum*, and, *Phyllostomus hastatum*.

Family (5) *Noctilionidae* : This family includes the Hare–lipped Bats. They have got very long legs and they feed on fishes.

Family (6) *Molossidae* : This family includes the mastiff bats. These are snub–nosed bats. They can walk with the help of fore arm when the wings are folded. *Eumops californicus* and *Tadarida teniotis* are the common examples of this family.

Family (7) *Desmodontidae* : This family includes the Blood–sucking Bats. They feed on blood by puncturing the skin of the victim with the help of sharp upper incisors. The stomach is reduced to slender tube. The blood is digested in the caecum which is very large. They are good walkers and they can run also. *Desmodus rotundus* and *Diphylla ecaudata* are common examples of this family. These bats are found in Mexico.

Aquatic Adaptations in Mammals

Adaptation to environment is one of the remarkable qualities of the living organisms. Adjustment of the animals to new environment is the process of adaptation and this is inevitable for the survival of the species. As the mammals are lung–breathers, they have become secondarily adapted to aquatic mode of life. Probably, severe competition for food was the main cause for some mammals which forced them to return to water. Aquatic adaptations have given rise to convergence.

General Trend of Aquatic Adaptations

1. *Stream–lined body* : The body is stream–lined or fish like in appearance in many aquatic mammals so as to offer least resistance during locomotion in water. The examples are the animals of *cetacea*, *sirenia*, and *pinnipedia*.
2. *Enlargement of size* : Aquatic mammals are normally larger in size because the energy is turned into growth force. For example, *Balaenoptera musculus* is several times bigger than the largest elephant.
3. *Modification of skull* : The cranium is short and wide. The facial portion is elongated and the zygomatic arch is reduced to vestige (e.g. cetacea).
4. *Shortening of Neck* : The neck is comparatively short. In whales all the seven cervical vertebrae are fused to form a solid and compressed mass of bone. Mobility of the neck is reduced. The chest becomes cylindrical.
5. *Simplification of Vertebrae* : The vertebral column is simple. The Zygapophyses become reduced, for the reason that

the body weight is supported by water. In cetaceans and sirenians, the sacrum is reduced. The diving force is transmitted longitudinally through the vertebral column.

6. *Light bones* : Lightness of bones is remarkable. The bones are light and spongy. There is trace of oil in the bones of cetacea.
7. *Lac of excrescences* : There is no trace of pinna or external ear (e.g. in whales) which would hamper swift locomotion in water. The external nares move towards the apex of the head in the case of whales, which can be closed whenever desired by the animal. The eye is also shifted higher is certain cases (e.g. *Hippopotamus*), in other cases it is adapted for aquatic vision. The caudal fin is generally horizontal. There is loss of hind limbs. The entire skeleton of fore–limb is enclosed in the skin.
8. *Locomotory paddles* : Fleshy and fan–like expansions of the body wall make their appearance in the aquatic mammals (e.g. whales). Dorsal and caudal fins may be present. Dorsal fin is present in Killer whale and absent in Balaena (Mysticeti). The caudal fin is horizontal, not vertical. The tail is divided into two parts. Tail propulsion is very important in whales as the hindlimbs are absent. The limb skeleton (forelimb) is enclosed in skin to form the paddles.
9. *Fatty layer, hairs and skin glands* : In the aquatic mammals a fatty layer known as *blubber* is present in the connective tissue below the skin. It retains bodily heat. The hairs are usually absent (e.g. whales and sirenians). The skin glands (e.g. oil glands) are absent as they have nothing to do with the aquatic mode of life. Salivary glands are also reduced.
10. *Mouth armament* : As the jaws are not used for mastication in most of the aquatic mammals, they have lost their power and as a result, the areas (e.g. coronoid process) for the muscle attachment are reduced. Teeth are also very simple and there is marked reduction in number (see Cetacea chapter). Whale–bone replaces the teeth in certain cases.
11. *Some other modifications* : There is well developed system of *retia mirabilia*. Greater obliquity of the diaphragm and its powerful musculature is also an adaptation (e.g. cetaceans).

Several mammals belonging to *Prototheria*, *Metatheria*, *Insectivora*, *Rodentia*, *Carnivora*, *Ungulata*, *Sirenia* and *Cetacea* are of aquatic habitat. Aquatic adaptations in some of the groups of animals have been described below :

(A) *Prototheria* : *Orinthorhynchus*, the Australian Water Mole, is an aquatic form. The legs are soft and paws webbed. The tail is broad and flattened. The external ear (pinna) is absent.

(B) *Metatheria* : The Water opposum, *Chironectes minimus* is the only marsupial adapted to an aquatic life. The hind feet are webbed and the tail is scaly and non–prehensile. The female can close the pouch at the time of diving.

(C) *Ungulata* (Artiodactyla) : The great African Hippopotamus, *Hippopotamus amphibius* is a common example of this order. It is more at home in water than on land. The skin is thickly lined with fat. It can swim easily in water. Nostrils, ears and eyes are placed high on the head. Thus, the hippopotamus can breathe and keep watch over the activities of other animals. The major part of the body is submerged for several minutes. There are very few hairs on the body. It dives well at night.

(D) *Rodentia* : The Muskrat is a good example of this order. These animals spend their life in ponds and rivers. They feed on aquatic vegetation. It has strong claws and incisors. The hind feet are webbed.

(E) *Cetacea* (e.g. whales) : The adaptations have been discussed separately later on.

(F) *Carnivora* (Sub–order — Pinnipedia) : They are marine carnivores and include walrus, sea–lion and true seals. The Atlantic Walrus, *Odobenus rosmarus* is found in Arctic Waters. The Pacific Walrus, *O. divergens* inhabits the Bering Sea. They do not possess external ears. The neck is quite distinct and palms and soles are naked. The walrus has only 18 teeth $\left(I\frac{1}{0}, C\frac{1}{1}, Pm\frac{3}{3}, M\frac{0}{0}\right)$. Both upper canines develop into downward–pointing tusks which may reach 27 inches in length. The forelimbs are primarily swimming organ. Hind limbs can be used for walking on land also. The clavicle is absent. Acromian and coracoid are short. Ulna is greatly expanded.

(G) *Sirenia* : This order includes the Sea–cows. Manatees (e.g. *Trichechus manatus*) and Dugong (*Dugong dugon*) are common examples. The skin is thick and hairless. The fatty layer (blubber) is prominent under the skin. The forelimbs are paddle like and the tail is flattened. There are only six cervical vertebrae and numerous vertebrae in trunk. However, manatees are less completely adapted to an aquatic life. They have scattered hairs, nails at the tips of flippers, and the tails are rounded. Dugongs are more aquatic in habitat like cetaceans. They are fish like in shape and the body is almost hairless. The forelimbs form the flippers and the hind limbs are present in the form of vestigial pelvis. The tail flukes are horizontal. There is no pinna. Nails are lacking in flippers. The tail is notched at the centre. In males, two upper incisors develop into tusks which are about 10 inches long. Clavicle is absent. The diaphragm is oblique and muscular. Molars are few in number showing signs of degeneration. Manatee (*T. manatus*) is commonly found along the coasts of the Gulf of Mexico and the West Indies, and Dugong (*Dugong dugon*) is widespread along the shores of Indian Ocean.

Conclusion

The impelling causes for adaptations are need for food and space, locomotion, and offense and defence. For these purposes both structural and functional adaptations are obvious. Such adaptations are called *convergent adaptations* because all are converged at one feature i.e. for aquatic habitat. It is in contrast to divergent adaptation (*or adaptive radiation*). Thus, the improvement of the efficiency of adaptation for a particular mode of life is considered as the process of evolution.

Dentition in Mammals

Teeth are present in all mammals with very few exceptions (e.g. some edentates and dolphins). These are arranged in a single line along the jaws. Only in the case of mammals the dentition has become sub–divided into fixed functional units : *incisors, canines, premolars,* and *molars.* Dentition is usually correlated with diet. In certain cases teeth do not occur in the adult condition. In *Tachyglossus,* when the milk teeth are shed, the permanent set of teeth never appears throughout life. A typical mammalian tooth consists of three kinds of tissue — *enamel*, *dentine* and *cement*. The enamel is endodermal in origin but the other constituents are dermal.

Parts of a Tooth and its Composition

A tooth consists of the following parts :

(i) *Crown* : The upper most part which projects above the gum.

(ii) *Root* : Part below the gum and embedded in socket or alveolus (is a thecodont tooth).

(iii) *Neck* : The portion where root and crown join.

Grinders have more than one root embedded in separate alveoli. A central pulp cavity is present in each tooth which has an opening at the apex of the root. If there are more roots, the opening is present at the apex of each root. The pulp cavity contains — connective tissue, blood vessels and also some nerves. In certain cases, roots are not formed and the crown portion continues to grow throughout life (e.g. incisors of rodents and elephants).

Enamel is the outer most part of the tooth which forms the shining surface and is the hardest tissue of the body. The dentine layer is present below the enamel. *Owen* has given the following composition of the dentine and enamel in the human tooth :

Chemicals	*Dentine(%)*	*Enamel(%)*
Calcium phosphate and Fluoride	66.72	89.82
Calcium Carbonate	3.36	4.37
Magnesium Phosphate	1.08	1.34
Other Salts	0.83	0.88
Organic matters	28.01	03.59

The cement substance binds the roots of the teeth to the neighbouring structures. A *periodontial membrane* of collagenous fibres is found between the root and socket.

Terminology and Succession of Teeth

Normally there are two distinct sets of teeth. The milk teeth (first set), and the permanent set of teeth. Sometimes only one set may be present, and accordingly *monophyodont* and *diphyodont* teeth are distinguished. When the teeth are uniform in appearance,

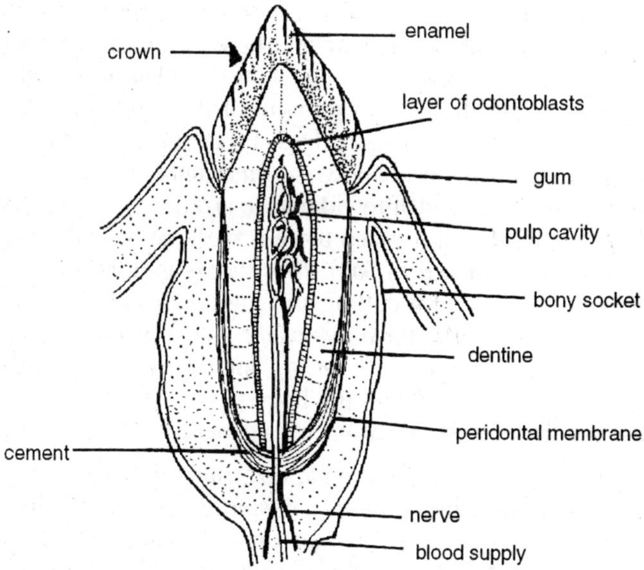

V.S. of tooth of Rabiit

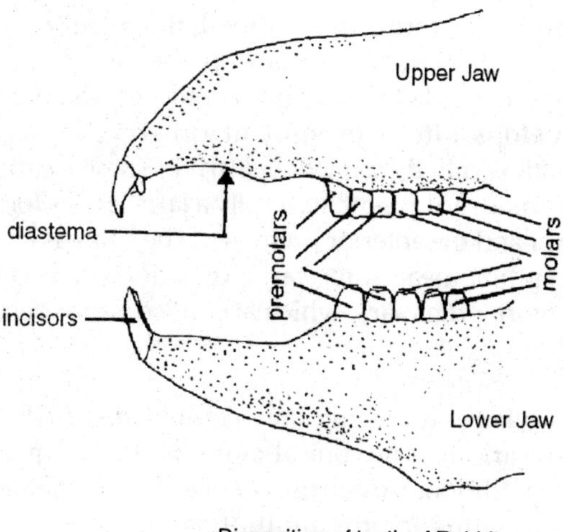

Disposition of teeth of Rabbit

Fig.39 : Teeth of Rabbit

it is known as *homodont* dentition (e.g. some dolphins), when it is divided into functional units, it is known as *heterodont* (e.g. man). In the latter case the teeth are of different shapes. The pre–molars and molars are known as cheek–teeth. The term *lophodont* is applied when the surface (crown) of these teeth are broad and they have sharp transverse ridges. When the molars have rounded or pointed cusps, it is known as *bunodont type* (e.g. man), when the cusps are fused into ridges, it is known as *lophodont type* (e.g. elephants), and when there is crescent–shaped of bands of enamel, it is known as *Solenodont type* (e.g. horses). In cats, the teeth are adapted for cutting and they are known as *Secodont*. In modern horses the term *hypsodont* is also used for continuous growth at a rate compensating for the wearing away of the grinding surface. When the horses possessed short - crowned molars the term *branchyodont* was generally used for that.

In the crown view, the molars are triangular (*triconodont*) except that each lower molar bears a posterior extension, known as the *tālonid*, at a lower level than the triangle, the *trigonid*. The triconodont tooth have three prominences in the crown. The middle one which is very prominent in the upper jaw is the *protocone*, having a smaller *paracone* in front and a *metacone* behind. In the lower jaw the corresponding terms *protoconid, paraconid* and *metaconid* are used. So in a tritubercular tooth the three cones are arranged in a triangle which alternate in the two jaws, the protocone on the inner side and the protoconid on the outside. When *talon* develops into a prominent tubercle (*hypocone, hypoconid*) the tooth is called *quadritubercular*. In some mammals (e.g. Rabbit) the canines are absent, naturally a large gap is formed between the incisor and the anterior premolar. The space is known as *diastema*. For cutting meat, some carnivores develop shearing edges on one or more premolars, which are called *carnassials*.

The Molar Pattern

There are two views, regarding how complicated molars of the mammals have arisen. The conical tooth has been supposed to be a primitive condition. According to one theory, the fusion of simple teeth accounts for the multiplication of roots and tubercles in various forms. According to another theory, parts have been developed on the primitive cone and as a result, triconodont shape was achieved. Then, these three cones have been shifted to the tritubercular position and later on, with the addition of hypocone, lophs and other parts modifications were brought about in various directions. The latter view has been supported by palaeontological records. In case of *Dugong* (order – Sirenia), a fusion of the enamel organs for the molar teeth has been reported.

Dentition in some important orders

Order 1 – *Marsupialia*

Nature of teeth and some important features have already been described under Metatheria. The dental formulae are given below in two cases only.

Kangaroo – $I\frac{3}{1}, C\frac{0}{0}, Pm\frac{2}{2}, M\frac{4}{4}$ = Total 32

American Opossum – $I\frac{5}{4}, C\frac{1}{1}, Pm\frac{3}{3}, M\frac{4}{4}$ = Total 50

Order 2 – *Insectivora*

The dentition is complete. All teeth are pointed or tuberculate. The total number of teeth may be 32, 36, 40 or 44. Some incisors and premolars disappear in certain cases. Some dental formulae have been given below :

European Shrew – $I\frac{3}{1}, C\frac{1}{1}, Pm\frac{3}{1}, M\frac{3}{3}$ = Total 32

Mole – $I\frac{3}{3}, C\frac{1}{1}, Pm\frac{4}{4}, M\frac{3}{3}$ = Total 44

Order 3 – *Chiroptera*

The dentition is complete. In the fruit bats the crowns of the molars are flattened. In the blood–sucking bats the teeth are comparatively few in number and they are slightly modified. The number varies from 20 to 38. The dental formula of a horseshoe bat is –

$I\frac{1}{2}, C\frac{1}{1}, Pm\frac{2}{3}, M\frac{3}{3}$ = Total 32

Order 4 – *Rodentia*

There is one pair of long incisors in each jaw and these teeth bear enamel only. The dentition is not complete as the canines are absent. The gap betwen the incisor and molars is known as *diastema*. In rats and squirrels the incisors are curved. Molars are less uniform in structure. The roots of these teeth are very long. The dental formulae of some members have been given below :

$$\text{Rat} - \quad I\frac{1}{1}, \ C\frac{0}{0}, Pm\frac{0}{0}, M\frac{3}{3} = \text{Total 16}$$

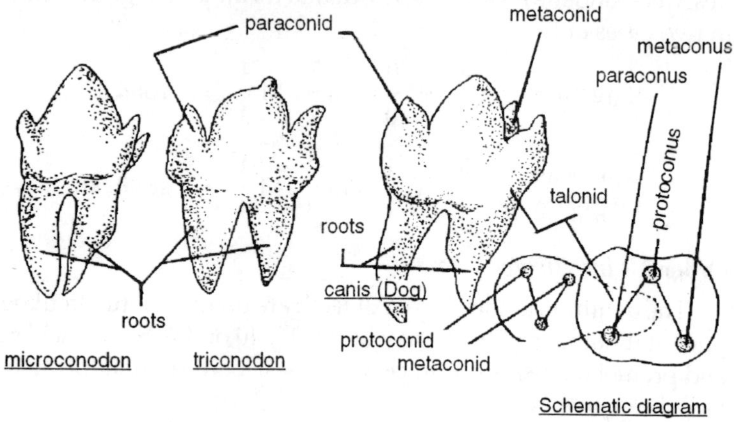

Fig. – Plan of molar teeth, left side in each case indicates anterior (after Osborn)

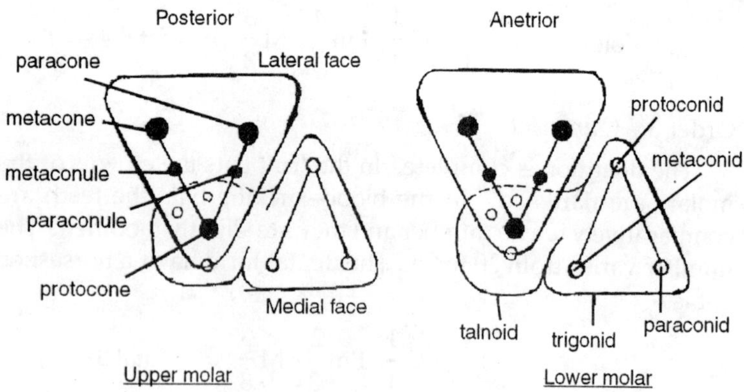

Fig.40 : Cusp patterns of upper (dark) molar teeth of primitive mammals showing the manner in which they fit when in contact (occlusion)

Squirrel – $I\frac{1}{1}, C\frac{0}{0}, Pm\frac{2}{1}, M\frac{3}{3}$ = Total 22

Guinea-Pig – $I\frac{1}{1}, C\frac{0}{0}, Pm\frac{1}{1}, M\frac{3}{3}$ = Total 20

Order 5 – *Lagomorpha*

Here are number of teeth is higher than the rodents described above. They possess an additional pair of incisors present inside the normal pair. The typical dental formula is :

$I\frac{2}{1}, C\frac{0}{0}, Pm\frac{3}{2}, M\frac{3}{3}$ = Total 28

Order 6 – *Carnivora*

They are hunting animals and as such, they possess very strong teeth, deeply rooted in the jaws. They have typically six upper and six lower incisors, and, fang like long canines. The last upper premolar and first lower molar are very much enlarged and have cutting edge. They are popularly known as *carnassial* – or flesh teeth. In the sub–order *Fisiipedia,* the cat bears only 30 teeth but the dog has 42 teeth. Dental formulae of cat and dog have been given below :

Cat – $I\frac{3}{3}, C\frac{1}{1}, Pm\frac{3}{2}, M\frac{1}{1}$ = Total 30

Dog – $I\frac{3}{3}, C\frac{1}{1}, Pm\frac{4}{4}, M\frac{2}{3}$ = Total 42

In the sub–order – *Pinnipedia,* incisors are normally present in both jaws, but the number is greater in upper jaw. The upper canines are fang–like. They may become tusks as in case of Walrus. The molars have crowns with cutting edges. The dental formulae of Eared seal and walrus have been given below :

Eared Seal – $I\frac{3}{2}, C\frac{1}{1}, Pm\frac{4}{4}, M\frac{1}{1}$ = Total 34

Walrus – $I\frac{0}{0}, C\frac{1}{1}, Pm\frac{3}{3}, M\frac{0}{0}$ = Total 16

Order 7 – *Edentata*

Molars are simple and numerous. They are without enamel or roots. Incisors are absent. The canines are also absent in the adults. Armadillo (*Dasypus novemcinctus*) of North America is an omnivorous creature and the food is caught by cylindrical tongue. In sloths there are eighteen teeth, five molars above on each side and four below which are rounded. In the Eocene Armadillo enamel cusps were persisting as evident from the fossil records.

Order 8 – *Proboscidea*

The dentition in elephants is slightly specialised. Lower incisors and canines are absent. The upper incisors grow to a great length particularly in the males. They may be straight or curved upwards. Enamel is present only on the tips. There are six molar teeth in each half of both upper and lower jaws but only one functional at a time. The lophodont teeth has become solenodont. The enamel folds are parallel in Indian Elephant but in the African Elephant the folds are lozenge–shaped.

Order 9 – *Sirenia*

Dentition is incomplete. Canines are absent and incisors are rudimentary. In the manatees the cheek teeth are covered with enamel and are bilophodont.

Order 10 – *Cetacea*

In toothed whales (Odontoceti), teeth are present in the lower jaw only. The total number of teeth is normally forty, each one about 8 inches long. Teeth are monophyodont and homodont in most of the cetaceans. The beaked whales have sometimes only one tooth in each half jaw. The narwhal has only two teeth in all only in the middle of the upper jaw. In male, one of these two teeth grows to six feet and forms a tusk. In the spermwhales functional teeth are present in the lower jaw only (about 25 on each side).

Order 11 – *Perissodactyla*

The teeth are either *lophodont* or *solenodon* type. The dentition may be complete (e.g. Horse) or incomplete (e.g. Rhinoceros). The dental formulae of Horse and Rhinoceros appear below :

$$\text{Horse} - \quad I\frac{3}{3},\ C\frac{1}{1},\ Pm\frac{3}{3},\ M\frac{3}{3} = \text{Total } 40$$

African Rhinoceros – $I\frac{0}{0}, C\frac{0}{0}, Pm\frac{4}{4}, M\frac{3}{3}$ = Total 28

Order 12 – *Artiodactyla*

The teeth are of *bunodont* type i.e. crowns of the molars have rounded tubercles. Tusk like canines are there. The dentition is normally complete, however, it may be incomplete in some forms.

Dental formula of Hippopotamus –

$$I\frac{2}{2}, C\frac{1}{1}, Pm\frac{4}{4}, M\frac{3}{3} = \text{Total } 40$$

Order 13 – *Primates*

In Primates, the dentition is complete. They have cutting incisors and fan–like canines. There is however variations in the structure of molars according to diet. Molars may have cutting crests for crushing the chitinous skeleton, and sometimes they have flat crowns with rounded tubercles. The common dental formulae are –

Lemurs and New world monkeys –

$$I\frac{2}{2}, C\frac{1}{1}, Pm\frac{3}{3}\ M\frac{3}{3} = \text{Total } 36$$

Old world monkeys, anthropoid apes and man –

$$I\frac{2}{2}, C\frac{1}{1}, Pm\frac{2}{2}, M\frac{3}{3} = \text{Total } 32$$

However, in Indris, the dental formula is –

$$I\frac{2}{1}, C\frac{1}{1}, Pm\frac{2}{2}, M\frac{3}{3} = \text{Total } 30$$

In *Tarsius spectrum* the formula is –

$$I\frac{2}{1}, C\frac{1}{1}, Pm\frac{3}{3}, M\frac{3}{3} = \text{Total } 34$$

Adaptive Radiation in Mammals

It is well known that in the "Struggle for existence" only most viable and adaptable species survive and others are eliminated in due course. Adaptation is the evolutionary divergence of animals from the primitive generalized type. The animals follow lines of adaptation in many directions and different modes of life in response to the type of particular environment. Reptiles and mammals exhibit the phenomenon of adaptive radiation in a very clear manner. During adaptive radiation the members of the same genus or closely related genera become adapted to different kinds of environments. As the animals invade different environments,

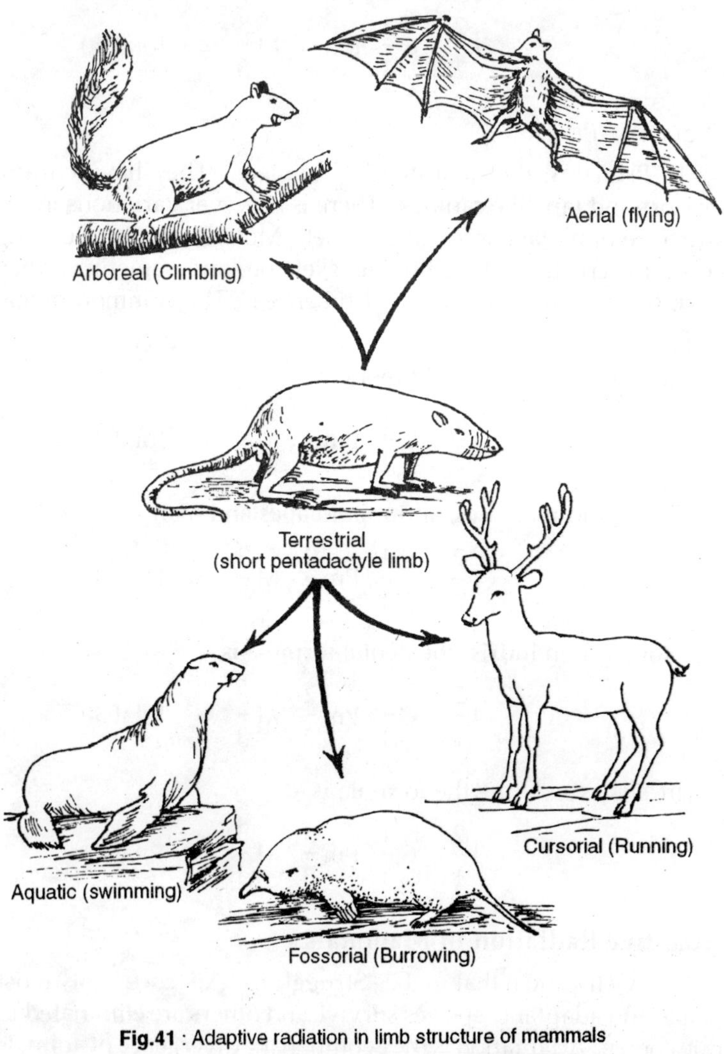

Fig.41 : Adaptive radiation in limb structure of mammals

they assume characteristic adaptations to meet their special needs. Thus, the animals enter the new adaptive zones and they look different from their original stock after a few generations. It is a case of divergence.

Thus *adaptive radiation is the evolution of animals in several specialized directions diverging from a common and generalized ancestral type.* Adaptive radiation may also be defined as the *entry of the organisms from a particular stock to new adaptive zones in response to their special needs.*

The limb structure of mammals provides a good example of adaptive radiation. The ancestry of mammals can be traced back to an insectivorous creature with pentadactyle limbs and short legs, walked with the soles of its flat feet. These were terrestrial creatures and are supposed to be the ancestors to modern mammals occupying five different habitats at present i.e. *Arboreal, Aerial, Cursorial, Fossorial* and *Aquatic*. The adaptations have been described in *'Adaptations'* – chapter of this book.

There are some impelling causes (food, safety and breeding ground) which force the animals to migrate to such habitats which are still unoccupied and there is practically no competition. Adaptive radiation is responsible for macroevolution and megaevolution. Adaptive radiation gives rise to new genera, species, orders, subclasses and also classes.

Tooth radiation in mammals is an important item for discussion in this connection. Teeth in mammals are divided into functional units. In *insectivorous* mammals, the premolars and molars are low crowned with simple cusps. In *carnivorous* forms they become high crowned having complicated cusps and shearing structures (*carnassial teeth*). Also, the canines are well developed. In *herbivorous* mammals the premolars and molars are either short crowned (*brachydont*) or high crowned (*hypsodont*). In toothed whales, teeth have become homodont, well adapted for grasping. In sperm-whales the teeth are absent. In some mammals, (*myrmecophagous forms*) the teeth have disappeared. Limb radiation in different groups of animals in response to particular mode of life, is also very important.

Chapter 8

Darwinism and Neo-Darwinism

Charles Darwin (1809-1882) was a British naturalist. In 1859 he published his observations and conclusions under the name *Origin of Species*. He proposed the theory of *Natural Selection*, popularly known as *Darwinism*. He held the view that new species originate as a result of Natural selection. Thus, Darwin's Natural Selection theory is a causo–mechanical explanation of the origin of new species. The theory is based on some observed facts and then a deduction from them. The fundamental propositions of Darwinism are the following :

1. *Over Production*: All organisms possess a potentiality of geometric increase. In majority of the cases, the number of offspring is several times greater than the number of parents. This tendency of multiplication is for the purpose of the continuity of the race. Thus, a large number of organisms are produced every year on the earth. Several examples can be cited. Oyster may produce 60-80 million eggs in a year. Similary, the ovaries of a cod fish contain 10 million eggs. The female *Salmon* produces about 28000000 eggs in a breeding season. The elephant is supposed to be a very slow breeder and has a life span of nearly 100 years. A single female gives birth to six young ones. It starts producing offspring at the age of 30 years and continues to breed till the age of 90 years. If it is presumed that all the progeny survive, then within a period of 750 years, 19000000 elephants will be produced from a single pair. Thus, there is enormous fertility in the organisms.

2. *Struggle for existence*: Thus, organisms multiply in geometric ratio but there is limitation of food and space. Naturally, there will be competition among the organisms

in order to get sufficient food and space. Every individual will try to beat down the other for getting the advantage of food and space. There will be a struggle among the organisms for their existence. The struggle for existence can be of three types :

(i) *Intraspecific Struggle*: It is the struggle between the members of the same species e.g. struggle of one *Felis tigris* with another *F. tigris* for their better survival.

(ii) *Interspecific Struggle*: It is the struggle between the members of different species e.g. the struggle between a *Felis tigris* and *Lepus cuniculus*. Here the former is the aggressor and the latter is the victim.

(iii) *Struggle with the environment*: It is the struggle of living organisms with extreme cold and heat, excess of moisture, drought conditions, storms, earth quakes, volcanic eruptions and other climatic conditions. Because of these factors it is difficult for all organisms to survive on the earth.

3. *Variations of Heredity*: No two individuals are exactly alike. There must be some differences between the two. These differences are known as *variations*. Variations give rise to new characters and heredity passes them to the subsequent generations. Development of new characters is necessary for the survival of the race. Some of the variations are temporary which are not inherited from one generation to another. But, there are some variations which appear in the progeny generation after generation. These are known as *heritable variations*. There are three sets of contrasting variations —

(a) (i) *Meristic variation*: Variation in number is known as meristic variation e.g. sometimes there are six digits in man instead of normal five.

(ii) *Substantive variation*: Some variations are substantive having independent existence, e.g. one family member may be very tall when others are dwarf.

(b) (i) *Continuous variation*: There are small changes which take place spontaneously in nature. It is a slow and gradual process. Darwin believed in continuous variation.

(ii) *Discontinuous variation*: These are large variations which lead to mutation, e.g. hornless cattle from the horned parents.

(c) (i) *Somatogenic variation*: Somatogenic variations appear due to changes in the somatic cells. These are acquired characters which appear due to influence of the environment. Dark complexion of human beings in tropical countries is the example of somatogenic variation.

(ii) *Blastogenic variation*: Blastogenic variations are changes due to modifications in the germ cells. These variations are inherited from one generation to another. Eye colour is an example of blastogenic variation which is heritable. For survival in nature, organisms must be capable of changing in response to the changing environment which includes food supply and climatic conditions mainly.

4. *Natural Selection or Survival of the fittest*: In nature, there is a constant struggle between the organisms. The beneficial variations are of great use in this struggle and they help the organisms in securing a better prospect for their survival. Only those organisms survive which have favourable variations. At least some of the variations are transmitted to the offspring. If, variations are not favourable the concerned organisms are weeded out. The process of selection always goes on and the organisms which have favourable variations and are most suitable from all view points are selected by nature for their survival. Darwin called it *Natural Selection* and Herbert Spencer called it *survival of the fittest*.

5. *Origin of new species*: There is continuous accumulation of divergences or variations in the organisms. A group of individuals belonging to a particular species show new adaptations. Continued variations in different directions and action of natural selection make the organisms different from their parents. The natural selection continues for many generations and the descendants become markedly distinct from their ancestors and they are graded as new species. Lion, tiger, leopard and cat have several features in common and as such it is

reasonable to say that they must have diverged from a common ancestor through variation and natural selection, and now they have been ranked as different species. Thus, the Darwinian concept advocates in favour of origin of new species by *Natural Selection.*

Sexual Selection

Sexual selection is '*Selection in Relation to Sex*' on which Darwin laid emphasis. He believed that there is competition for mating also. It is a fact that there is struggle among the males for possession of female in certain animals and the choice of male is exercised by the female. The female selects a male for mating who has some distinctive characters like suitable body size and morphological appearance, brilliant colours, conspicuous ornamentation and pleasing behaviour such as dancing, singing and effective courtship techniques. So, this is known as sexual selection.

Kellogg has pointed out some difficulties in accepting this theory. According to this author, the sexual selection can operate only when the number of males is much greater than that of the females, or there is polygamy. Also more progeny can not be produced as a result of sexual selection. In certain species the female may prove to be passive in the matter of sexual display. Attractive colouration and ornamentation, and behavioural distinctions are insignificant. The theory has not been described in terms of reproductive fitness.

The sexual selection theory, though it may not be adequate, but it is never wrong. The Secondary Sexual characters are of considerable importance.

Pangenesis Theory

The *Theory of Pangenesis* was proposed by Darwin to explain the inheritance of characters. But, this theory has become outdated in the light of present knowledge of gene structure. This theory can be illustrated in the following points :

1. All the somatic cells of the body produce minute particles, the *pangenes* or *gemmules* which are carried away by the blood stream and are concentrated in the germ cells of both sexes. Pangénes are produced at all stages of development.

2. The germ cell (gamete) represents a minute replica of the body of the parent and these are capable of developing into the same kind of body (adult) with minute details.
3. Each pangene in the developing individual regulates the development of the organ from which it originated from the parent.
4. Sometimes certain pangenes are in dormant condition for several generations and then they develop. So, there is appearance of characteristics in the offspring which were borne by their ancestors (atavism) and are not seen in the parents.

Later on, this theory was totally discarded when *Germplasm theory* of Weismann was proposed.

Neo–Darwinism

Darwin's theory of Natural Selection was accepted for a long time but later on, several objections were raised against this causo–mechanical explanation of the origin of new species. Some of the important objections are —

1. Darwin could not distinguish between heritable and non–heritable variations, also, he did not explain the origin of variation.
2. The Natural Selection can not be considered as a sole cause for the origin of new species but the new species originate as a result of mutation.
3. The Natural Selection does not explain the effect of use and disuse and the presence of vestigial organs.
4. He has described the survival of the fittest only but not the arrival of the fittest in any form.
5. Darwin believed in *Blending inheritance* i.e. the offspring should resemble their parents. He had no idea of *Particulate* inheritance.
6. He believed in small and fluctuating continuous variations.
7. The *Pangenesis theory* was criticised.
8. Several differences between species are not of adaptive value.
9. No explanation for over specialization was given, as in dinosaurs.

10. The sexual selection theory proposed by Darwin was considered inadequate.

On the basis of objections raised by Bateson, Hogben and some others the death of Darwinism from the Biological laboratory was proclaimed. But, all the objections are not considered very serious at present and Darwinism is supposed to be very much alive even today. Actually, the objections are based upon lack of full knowledge of what Darwin wrote. Some of the objections are largely academic and do no stand up valid under actual test. Some biologists of international repute have supported the facts related to Natural Selection theory. Some of the opinions have been given below :

Prof. V.L.Kellogg in this book *"Darwinism Today"* has expressed a sense of defence for Darwinism. Mendelians observed that mutations could be either continuous or discontinuous. New data from different disciplines i.e. cytology, genetics, experimental embryology, paleontology, Ecology and Systematics were obtained and new approaches to evolutionary problems have been made. Selection has been studied in the nature as well as in the laboratory by applying quantitative and qualitative methods.

Blending inheritance was also shown by R.A.Fisher (1920). T.H.Morgan a critic of Darwinism became the upholder of Darwin's ideas. Prof. C.C.Nutting has also expressed his view and considered Natural Selection as an important factor in evolution and quite in harmony with the mutation theory and Mendelism. Now it has been argued that the theory of Pangenesis was a working hypothesis which was simply a starting point for further investigation. The Darwinism, thus reborn, is a modified Darwinism. It gives a naturalistic pattern of evolution. According to geneticists and germplasm hypothesis with the aid of mutation and mechanism of Mendelian inheritance provide the necessary explanation of the evolutionary causes. But mutation does not furnish any account for adaptation as revealed by Weismann. It is only the Natural Selection which gives satisfactory account for adaptation.

Some worth mentioning facts regarding Neo–Darwin concept have been described below :

1. *Directional Selection*: It may be called a progressive selection. Here, shift of population takes place. Directional selection favours those that vary in one general

direction. Those which favour opposite direction are eliminated. This type of selection produces evolutionary changes. Development of molar teeth in horses is an example of directional selection.

2. *Stabilizing Selection*: It is also known as normalizing selection. It operates when the environment is not changing. This selection eliminates the deleterious mutant gene and gene combination. This stabilizes and protects the normal development pattern of the species. It can stop any change from normal trait.
3. *Dynamic Selection*: It permits the species to maintain itself in the changing environment. The species can take advantage of new ecological opportunities. But, it requires favourable genotypes otherwise, the survival of the species becomes difficult.
4. *Differential Reproduction*: It means the production of more progeny of individuals of a particular genotype in comparison to other individuals having different genotype. Unfavorable genes cannot spread in the population. There is rapid spreading of genes producing advantageous characters. The disadvantageous gene combinations and mutations are eliminated. New species can be given rise, being adapted in new localities.
5. *Genetic drift*: Genetic drift brings about random changes in the gene frequency. Genetic drift is a consequence of drawing a small (random) sample of gametes for the formation of next generation. Genetic drift can fix new alleles of genes that arise as a result of mutation from time to time and eliminate the original gene. Thus the genetic make up of a small population can be changed. It also reduces genetic variability in the populations by eliminating old or new allele.
6. *Isolation*: Division of a single population into two or more groups with some barriers to interbreeding is known as isolation. Besides geographical isolation, genetic isolation is also essential for the origin of new species.
7. *Neoteny*: In this phenomenon, the sexual maturity occurs at the larval stage of the animal. It is considered as one of the forces which brings about evolutionary changes.

8. *Diverging Selection*: In this type of Natural Selection, the amount of variance within the population is increased. It acts to breakup a homogenous population into several different adaptive norms.

To sum up, Natural Selection and mutation together constitute the raw materials for the origin of new species. The mechanism of Mendelian inheritance also explains the progressive mechanism of evolution alongwith natural selection and mutation. Mutation provides bulk of evolutionary materials and useful gene combinations. It facilitates favourable variations. Isolation and Genetic drift also assist in the formation of new species. The population moves to a superior adaptation. The whole process of evolution discussed above has been called *Neo–Darwinism* and the supporters of this view are known as Neo–Darwinians.

Chapter 9

Speciation and Species Concept

Speciation means the formation of new species. During the process of speciation the production of new gene complexes takes place which are capable of ecological shifts. Species is the lowest rank (unit) in the classification of animals. It has also been defined as a group of genetically identical individuals. As a matter of fact, several definitions of species have been given by different authorities. Species are closed genetic system i.e. they are protected against gene flow from another species. According to Dobzhansky (1950) "species are Mendelian population which share in common gene pool". Mayr (1942) stated that "Species is a group of actually or potentially interbreeding population which are reproductively isolated from such groups". Species can interbreed among themselves but they are reproductively isolated from other populations.

Formation of Species (Speciation)

New species are formed by two ways —

1. *Phyletic Speciation* : When one species is transformed into another in due course of time the process is known as phyletic speciation. Only one species exists at a time. For example species A is transformed into B and B is transformed into C and so on. This method of speciation is of two types —

   (i) *Autogenous transformation* : This transformation takes place when evolutionary agents like Natural Selection, Mutation, Genetic drift and Migration operate.

   (ii) *Allogenous transformation* : In this case, new species are formed by hybridization mainly.

2. *True speciation* : Here new species are formed by multiplication of original species. Thus, splitting or fragmentation gives rise to new species. Thus, two or more species co–exist with the parental species. True speciation can be divided into two types —

   (i) *Instantaneous Speciation* : This term was suggested by Mayr (1949, 1963). Here one species splits into two species suddenly in a single stroke by the development of chromosomal aberration.

   (ii) *Gradual Speciation* : Here the species splits into two species gradually, by the accumulation of genetic divergence and reproductive isolation. Geographically isolated populations (allopatric population) may give rise to new species because of genetic divergence. Two neighbouring population (sympatric populations) form separate species.

Sibling Species

This term was used by Mayr (1942). Morphologically similar and reproductively isolated species are called Sibling species. Thus, they are sympatric species. These populations are called races. Several species of Drosophila are Sibling species e.g. *D. pseudoobscura* and *D. persimilis*.

Infra Specific Categories

Animal populations have been conveniently divided into small or infraspecific categories below the species level e.g. *Sub–species, Clines, demes* etc.

Demes

It is the basic unit of animal populations and it is defined as a community of interbreeding individuals of a species at a given locality which share in a common gene pool. The demes live in areas where the conditions are most favourable for them. Thus, a particular species is organised into demes because of unsuitable circumstances. A deme is isolated from its neighbours and live independently. Two categories above the deme are *sub–species* and *clines*.

Sub–species

A sub–species is an aggregate of geographically isolated demes of a species. They differ both in morphology, and genetic composition from other population of the species. They can interbreed and produce fertile hybrids. Sub–species occupy separate territory but they are not reproductively isolated. It is considered as a small step in the divergence of various species from a parental species. A species that contains two or more sub–species are known as *Polytypic species*. The *monotypic* species is not divided into sub–species.

Clines

It is a group of local populations (demes) exhibiting regular and gradual stepwise modifications from one part of their geographic range to another. Species are not always organised to form sub–species. Characters of one deme vary from another throughout the range of species. Huxley used the term demes for such a group of local population. Changes in the characters of clines may be constant (*Internal cline*) or may not be constant (*Intergroup* or *stepped clines*).

Patterns of Speciation

There are four important pattern of speciation which have been described in short below :

1. *Allopatric Speciation*: Allopatric populations are those populations of species or races which inhabit different areas i.e. they are geographically isolated. Here interbreeding is possible. The evolution of allopatric populations into separate species is called allopatric speciation. Geographic isolation is the first step in speciation. The geographical barriers separate the original populations into allopatric populations. In the beginning the gene pool of each allopatric population is similar. Owing to the presence of barriers, there is no possibility for the mixing up of genes between these populations. The factors like Mutation, Natural selection, Genetic drift and Recombination act on these populations. As a result, genetic divergences are developed by these populations. When these populations are brought into the same territory, they fail to interbreed and thus, reproductive isolation comes into existence. Now, these populations are graded as new species.

2. *Sympatric Speciation*: Sympatric populations are those which inhabit the same geographical areas. Here

interbreeding is not possible. The evolution of sympatric populations into separate species is called sympatric speciation. Evolutionary changes go on in different populations. Because of residing in the same locality there is competition between the various populations with regard to food, space and other requirements. Behaviour of population involved in mate selection, food selection, selection of habitat and other necessities have to do much in giving rise to variations. Morphological form, behaviour of the species and other characters form an adaptive complex. At the same time, the factors like Mutation, Recombination, Polyploidy, Hybridization, Natural Selection, Genetic drift, Isolation and Founder Principle etc. also operate in the populations. Naturally genetic divergences appear in the populations which had once the similar gene pools. After several generations, the genetic divergences are established. These populations when brought in contact fail to interbreed and thus, gene flow is prevented. Each population now maintains a separate and distinctive gene pool showing reproductive isolation. So, they are ranked as new species.

3. *Quantum Speciation*: Quantum speciation is a substantial speciation as stated by Ayala (1975). Several examples among Drosophila species inhabiting Hawaii islands have been cited by H.L.Carson. In Quantum speciation, higher taxonomic groups like orders and classes evolve. It is a more rapid and more abrupt mode of species formation. It involves sudden change in the organization of population to a new equilibrium. In this rapid speciation, sub–species are not formed. Genetic drift plays an important role here.

4. *Parapatric Speciation*: Geographical isolation is not involved in Parapatric speciation. Natural selection gives rise to reproductive isolation, which is the main operative force here. It is an intermediate way of speciation between allopatric and sympatric methods of speciation. Here divergence takes places between continuous populations. It is also a rapid process but involves very few members of the original (Parent) populations. Bush (1975) stated that this speciation may occur "whenever species evolve as continuous populations in continuous clines".

Factors Influencing Speciation

It has already been mentioned that speciation is the production of new gene complexes. Changes are produced in the genetic composition of the individuals, which are nothing but genetic divergences or variations. These are produced by several factors which act on the populations. These factors are —

(1) Mutation, (2) Natural Selection, (3) Genetic drift, (4) Recombination, (5) Polyploidy, (6) Hybridization, (7) Isolation, (8) Founder Principle.**

** These factors have been described in various chapters of this book and hence not repeated here, readers should see other chapters too.

Chapter 10

Modern Synthetic Theory

The modern synthetic theory is based on the accumulation of facts and theoretical conclusion from a number of scientists like, Woodward (1864-1944), Wright (1931), Watson (1866-1951), Marsh (1831-1899), de Vries (1840-1935), Cope (1840-1897), Dobzhansky (1937), Julian Huxley (1942), Watson and Crick (1953), Rensch (1960), Mayr (1942, 1943, 1963) and others. The synthetic theory provides modern concept of evolution and explains the mechanisms by which evolution takes place. Several facts have been introduced and the idea of evolution has been made conceivable. Some authors have attached this theory with Neo–Darwinism but, there is no justification in doing so. This theory deserves independent status.

This theory has been constructed on the following concepts which can explain the actual mechanism of evolution. These facts are —

1. Gene Mutation
2. Chromosomal Mutation
3. Genetic recombination
4. Hybridization
5. Natural Selection
6. Genetic drift
7. Founders principle
8. Isolation

Now, these facts have been described below in short.

1. *Gene Mutation*

According to *de Vries*, mutation meant a striking change of phenotype. It is the inception of the hereditary variation. Change in gene is known as point mutation. As the gene is composed of DNA, mutation is due to a change in the gene or DNA. The DNA molecule has a coded information in the form of a particular sequence of its base pairs. A change in the sequence of arrangement of *Pyrimidines* and *Purines* changes the functions of these genes causing a change in the phenotype. So, the gene mutation is the result of an alteration in the sequence of Purine, Pyrimidine pairs in a DNA molecule. Watson and Crick have suggested that mistakes in pairing might occur at the time of duplication of DNA. Mistakes in pairing leads to mutation. In certain cases, the analogue can take the place of natural substances in the same biological reaction. For example – *5–Bromouracil* can replace thymine and the former is mutagenic. It is incorporated into the DNA. Sometimes *Adenine–thymine* is replaced by *Guanine–Cytosin* under certain circumstances and causes changes in the gene by changing base pair. There are several mutagens which change the bases of DNA molecule. For example, *Adenine* can be changed to *Hypoxanthine* which constitutes a mutation.

There is permanent alterations in base sequences which may include deletion of bases, insertion of bases and inversion of bases. They involve breakage and reunion of DNA segments. Replacement of base–pair takes place during replication by transitions and transversions. There is no breakage of DNA in this case.

2. *Chromosomal Mutation*

Mutation is also brought about by certain changes in the chromosomes. Some important causes have been described below:

(i) *Chromosomal aberration* — There is change in the arrangement of chromosomal segments.

(ii) *Inversion* — Here, intra–chromosomal segmental arrangement takes place.

(iii) *Duplication* — Duplication of chromosomal segments takes place.

(iv) *Translocation* — Here, interchromosomal segmental arrangement takes place.

(v) *Deletion* — Sometimes, there is loss of a segment of chromosomes.

(vi) Changes in the number of chromosomes also takes place. This can be discussed under —

(a) *Polyploidy* — Increase in the number of chromosomes takes place.

(b) *Heteropolyploidy* — Addition or subtraction of less than a whole set of chromosomes takes place.

(c) *Haploidy* — There is decrease in the number of chromosomes.

3. *Genetic recombination*

Recombination is also an important source of genetic variation and adds in the production of evolutionary materials. New genotypes can be produced from the previously existing types. It means that the formation of new gene combinations appear which were not present in the parents. Mixing of a particular *allele* with a series of genes which were not in their association earlier takes place during crossing over (Meiosis). Thus new combinations of genes are produced. According to Moody (1970) the process of recombination involves reassortment and recombining of genes already present in the parental form. Random union of gamets at the time of fertilization, chromosomal aberration and interbreeding also add during recombination of genes. All these processes cause position effect which facilitates the formation of new genotypes. Beneficial combinations are always favoured by Natural Selection as suggested by Moody (1970).

4. *Hybridization*

Interbreeding between the members of different races, species and also among some related genera is attempted and the resulting offspring are called hybrids. The process is known as hybridization. It causes genetic variation in the population. New characters are developed and the new individuals are reproductively isolated from the parental populations. Thus a new species may be formed. The process of hybridization leads to *Sympatric Speciation.* The new genotypes adapt themselves to the changed environment.

5. *Natural Selection*

Darwin's Natural Selection theory is a causo–mechanical explanation of the origin of new species. This theory is based on some observed facts and then a deduction from them. Various propositions in sequence are : Over production, Struggle for existence, Variation and heredity, Survival of the fittest and then the final conclusion was the *Origin of new Species*. Thus, according to Darwin the new species originate as a result of *Natural Selection*. Several objections were raised against the Natural Selection theory. Also, *Mutation theory* was proposed by *de Vries*. The death of Darwinism was proclaimed from the biological laboratory. Later on, the importance of Natural Selection theory was realised by several workers like Prof. V.L.Kellogg, R.A.Fisher, C.C.Nutting, T.H.Morgan and others. According to them, *Natural Selection* is still an important factor in evolution quite in harmony with both Mutation theory and Mendelism. Natural Selection, Mutation and Mechanism of Mendelian inheritance furnish all the necessary explanation for the causes of evolution and origin of new species. Directional selection, Stabilising selection, Dynamic selection and Differential reproduction are some of the important factors associated with Natural selection which were taken into consideration. It is only Darwinism which explains the process of adaptation.

6. *Genetic Drift*

It is also an evolutionary force which operates in small population. It was explained by Sewalt Wright (1931). Genetic drift produces variations in small population. Changes are brought about in gene frequency just by chance. Random changes in the gene frequency are said to be due to genetic drift. The mutation produces as a result of genetic drift is either lost or fixed. The operation of genetic drift helps in the origin of new species.

Suppose that two alleles of a gene A and a are equally frequent in a small population. The frequency in a certain generation q = 0.5. In the subsequent generation the frequency will fluctuate up and down from 0.5. Wright (1932, 1948, 1949) showed that the frequency may lead to important results. So long as the value of q remains between 0 to 1, the fluctuations are reversible but when the latter values are reached an irreversible change takes place. One of the two alleles is lost and other has reached a state of fixation and the population thus becomes homozygous.

$$\text{Heterozyous} \xrightarrow{\text{Genetic drift}} \text{Homozygous}$$

Thus, genetic drift can preserve or destroy genes without distinction. Particularly in small populations, the Hardy–Weinberg equilibrium is upset by genetic drift.

7. *Founder's Principle*

This principles is similar to genetic drift. It can also bring about genetic divergence in certain populations. Founder's Principle was proposed by Mayr (1963) and Sheppard (1960). When a new population is established during the course of isolation, its gene pool is not identical with that of the parent population because of *sampling error.* The two populations are subjected to evolutionary pressures independently and the difference is further improved and it leads to genetic divergence. For example, suppose that a small group of animals from a continent succeeds in colonising an island. There will be a change in the gene frequency as a result of Natural Selection. A new gene pool will be evolved which may be quite different from the original species. This is deemed to be an evolutionary change. This new population on the island becomes adjusted to this new environment. Selective forces are different in the two population leading to change in the gene frequency.

8. *Isolation*

Isolation is an important factor in evolution and it plays a significant role in the origin of new species. "The division of single population into two or more groups due to some barrier to interbreeding is called isolation". There are several processes by which two related population living in the same area can remain distinct. These have been called "isolating mechanisms". The Principal, isolating mechanisms are (a) Geographical isolation, (b) Ecological isolation, (c) Psychological or Sexual isolation, (d) Mechanical isolation, (e) Physiological or Reproductive isolation, (f) Hybrid Inviability, (g) Hybrid Sterility, (h) Hybrid break down.

Isolating mechanisms operate under the control of genetic and chromosomal factors. Accumulation of genetic changes gives rise to new species. Natural Selection and Mutation Constitute the raw materials for the origin of new species. Genetically conditioned isolating mechanisms result in reproductive isolation. The gene exchange between species is restricted or suppressed in the sympatric population.

(Isolating mechanisms have been discussed in detail in 'Isolation' Chapter).

Conclusion

Thus, two populations becomes separated from each other by

Geographical or environmental barriers. They are acted upon by forces like *Natural Selection, Mutation* and *Genetic drift* etc. Bulk of evolutionary material are provided, favourable variations are produced and also useful gene combinations are provided. Genetic diversity results between the two populations. Finally the *allopatric* populations become differentiated and gives rise to reproductive isolation. When the two populations come into contact they remain distinct and are said to be *sympatric*. So, new species are gradually formed.

Chapter 11

Isolation and its Role in Evolution

Isolation is the segregation of populations by some barriers which prevent interbreeding. Because of isolation, gene flow between populations is also prevented. Isolation is an important factor in organic evolution. It plays an important role in the origin of new species. Each population develops genetic divergence independently. Modern evolutionists attach much importance to isolation particularly when sub–species are changing into species otherwise there is possibility that the process of evolution will be delayed.

The division of a single population into two or more groups due to some barrier to interbreeding is called isolation. There are several processes by which two related populations living in the same area can remain distinct. The agency which brings about isolation is called isolating mechanism by Dobzhansky. Exchange of genetic material is either prevented or greatly reduced.

As long as the related gropus remain geographically separated from each other, they are referred to as *Allopatric species*. But, if they move into the same territory and fail to interbreed with the formation of intermediates, they are referred to as *Sympatric species*.

Classifications and examples of isolating mechanisms have been given by Durietz (1930), Fisher (1930), Dobzhansky (1937), Huxley (1942), Muller (1942), Patterson (1942), Mayr (1942, 1947, 1948), Stebbins (1950) and others. The principal isolating mechanisms have been described below :

1. *Geographical isolation*: The populations occur in the different territories, either within a continuously inhabited area or separated by distributional gap. They are prevented from interbreeding by the presence of certain

barriers like large bodies of waters, high mountain ranges, deserts and forests, extreme of temperature etc. Because of barriers, two populations are completely out of touch with each other genetically. As a result new mutations, genetic drift and action of natural selection in one population have no effect on the other population. Such a geographical barrier exists between the fauna of an oceanic island and that of the main land.

2. *Ecological or Seasonal*: Ecological isolation means isolation by habitat (Moore, 1949). Populations living under different environmental conditions remain isolated from one another and are prevented from interbreeding. Environmental isolation depends upon differences in food habits and other physiological requirements of the animal.

   The stickle–back fish (*Gasterosteus*) occurs in Belgium. One of them lives in freshwaters but the other lives in Sea water and migrates to estuaries in spring and summer seasons where it breeds. The population differ in the average of certain morphological traits. The freshwater form has an osmoregulatory capacity for maintaining the chloride content of its blood in the water of low salinity. The migratory form losses chlorine ions in the water of low salt concentrations. This leads to death of the fish within a very few days. Hybrids can be easily produced by artificial insemination of the eggs but in the natural habits it is never possible.

   Another example is that of two species of *Bufo* whose breeding seasons do no coincide and as such, hybrids can not be produced in the natural habitats. Fertile hybrids can be produced by crossing *Bufo americanus* with *Bufo fowleri* in the laboratory. In the natural habitat, *Bufo americanus* breeds early in the season but *B. fowleri* breeds late and hence they remain distinct in nature.

3. *Psychological or Sexual or Ethological Isolation*: Among certain populations, interbreeding is possible, but due to mating preferences, say, lack of attraction and other factors, it is prevented. In many insects and birds, copulation is preceded by an elaborate courtship and closely related species differ in the details of this process. In some families of insects, males recognise their females

by sight, in others the males must both see and touch the females. In some species visual stimulus plays no part and sense of touch is paramount.

Males usually approach females of foreign species but they are rebuffed either immediately or after some time when postured. According to Mayr (1950) males of *Drosophila pseudoobscura* and *D. persimilis* court females of either species at random. They are rejected by foreign females more often than by conspecific one as revealed by this worker. The choice is exercised by the female and not by the male. Several birds show beautiful plumage and the females exercise the choice of mates.

4. *Mechanical Isolation*: The structure of genitilia is complex specially in insects. The "Lock and Key" theory was propounded by Leon Durfour in the Pre–Darwinian days and later on it was elaborated by K. Jordan (1905). This theory states that the female and the male genitalia are so exactly fitted to each other that even a little deviation in the structure of either makes the copulation practically impossible. Different species are isolated from each other by the non–correspondence of their genitalia. In several species of Drosophila, the interbreeding may take place but due to non–correspondence of genitalia it is not feasible.

5. *Physiological and Reproductive Isolation*: It has been experimentally shown by Patterson (1942, 1947) that in case of some interspecific matings in Drosophila, the sperms fail to survive in the receptacles of the females of other species. However, in conspecific females the mobility is retained for a longer period. Thus there is insemination reaction in certain species following copulation as shown by Patterson (1946, 1947). Following copulation, there is immediate secretion of fluid in the cavity of vagina which causes swelling of the organ which may persist for several hours after interspecific copulation. Insemination reaction is violent particularly when the copulation is attempted by a male of foreign species. Swelling of vagina continues for several days in such cases. Not only that, the secretion in some cases solidifies, and the passage of eggs is obstructed and, thus, the female in made sterile.

If there is genotypically conditioned differences between two populations, the exchange of genetic materials is either completely restricted or at least suppressed. Because of intrinsic and genetically conditioned properties, sympatric sexual populations can be maintained distinct. According to Mayr (1942), the genetically conditioned isolating mechanisms result in reproductive isolation. Dobzhansky holds the view that the reproductive isolation between sexually reproducing and cross fertilizing species is caused by changes in the complementary gene complexes. Natural selection brings about reproductive isolation between closely related populations.

6. *Hybrid Inviability*: The hybrid zygotes are inviable and its development is irregular. They are adaptively inferior to those of the parental species. The life of the hybrid zygote may be cut short at any stage before birth. The process of development is rarely normal. Several disturbances may occur in the zygotes e.g. chromosome elimination during cleavage, arrest of gastrulation or of organ formation to death of the embryos before the formation of hybrid progeny — all being common features. Warwick and Berry (1949) reported that sheep and goat hybrids are found to develop normally in early stage of development but they die much before birth. The isolating mechanism has been studies by Moore (1946, 1949) in a group of North American species of frog (*Rana*) and it was revealed that mostly the hybrids die before birth.

7. *Hybrid Sterility*: Hybrids are generally sterile or semisterile and they fail to produce a normal complement of functional sex cells. There is no exchange of genes. Hybrid sterility has been described in *Drosophila* by Patterson and Stone (1952) giving a complete account. The mule is a good example of hybrid sterility. In the sterile hybrids either the sex organs fail to achieve normal development or abnormalities in meiosis prevent the formation of normal gamets. Federley (1913) discovered that the chromosomes fail to form bivalents at meiosis in the sterile or semisterile hybrids between *Pyaera anachoreta, P. curtula* and *P. pigra*. Fused spindles and failure of cell division were observed in the spermatocytes, and degeneration at

the spermatid stage was also marked. In the hybrids of *Drosophila melanogaster* and *D. simulans*, Kertis (1933) reported rudimentary gonads and development does not proceed further in spermatogonia and oogonia. This also happenes in the hybrids of horse and zebra.

8. *Hybrid Breakdown*: In certain cases, hybrids may be vigorous as the parental species, particularly in the F1 generation. In some cases they are only partly fertile. But, they give rise to weak and sterile progeny in the second generation. They are poorly viable or lethal. This is known as hybrid breakdown. The general viability of the back cross progenies is however low in comparison with the parental species and with F_1 hybrids.

Origin of Isolating Mechanism

As a result of phylogenetic change in gene functions, the reproductive isolation arises. There is an integrated system of genes in a functional genotype. Reproductive isolation is due to differences in the genes of allopatric population when the formations of species and sub–species initiate. Accumulation of gene differences goes on and they are the essential constituents of the genotype. This idea has been proposed by Muller (1939, 1940, 1942, 1950). Further, this author believes that reproductive isolation is due to differences in the genes which arise in the process of the origin of subspecies and species in a population.

However, Dobzhansky (1940) believes that physiological isolating mechanisms may be a product of natural selection. This view was also proposed by Fisher (1930). Natural selection brings about reproductive isolation between closely related populations. Sterile or poorly adapted hybrids are eliminated by natural selection. Thus, reproductive potentiality of the interbreeding population is gradually decreased. Natural selection acts against hybridization.

Genetic diversity is the main cause for the origin of new species and some geographic or environmental isolation is essential in the beginning to produce genetic differences.

Isolation and Natural Selection

Isolation is an important factor in evolution. It is as important as Natural Selection. As a isolation, a single population is divided into smaller groups. The new isolated group has to adjust itself in a new environment unlike parent stock. Now, natural selection preserves their advantageous genotypic combinations. Thus it preserves only the beneficial ones and the rest are eliminated. The genetic differences between the isolated group and the parent species increase. So, isolation assists selection in the formation of new species.

Role of Isolation in Speciation

The geographically separated populations are called allopatric populations. According to Mayr (1963) the geographical isolation is a necessary step in the development of diversity between populations which later on contributes in the origin of new species. The first step, in the development of diversity between populations which later on contributes in the origin of new species. The first step, in the multiplication of species, is a physical separation of a part of the species which goes on its own way genetically. Isolation depends on the genetic contents of the isolated population. Isolated populations are exposed to all kinds of environmental factors. When isolation is not complete for some reasons or other, the gene pool in the different populations will not remain completely separated. In such cases neighbouring populations may interbreed. The rate of gene flow depends on the extent of isolation. When the isolation is complete, the gene pools are out of touch in the two or more populations and in that case the two populations are free to go their evolutionary ways. The natural forces like mutation, natural selection, recombination and genetic drift etc. act upon the isolated populations. The elemental forces leads to progressive genetic divergence in the separated population. The genetic divergences finally facilitate the development of reproductive isolation. When the reproductive isolation becomes complete, the isolated populations fail to interbreed, even if they are brought in contact with each other. Thus isolation brings about allopatric speciation but subsequently reproductive isolation becomes complete and sympatric speciation stage is reached. Thus new species are formed.

Chapter 12

Lamarckism and Neo–Lamarckism

Jean Baptiste De Lamarck (1744-1829) was a French naturalist. His evolutionary doctrine was published in *Philosophic Zoologique* in the year 1809. The main idea of this theory lies in the fact that "acquired characters in the organisms are inherited to the subsequent generations". Thus, it states that evolution of new types of organisms takes place by the effect of environment which causes changes in the shape and organisation of animals and these changes acquired, are passed on to the offspring. There are four important propositions of Lamarck's Theory (Lamarckism) :

1. Living organisms tend to increase in size continuously.
2. Formation of a new organ in the body of the animals is the result of new, growing needs.
3. If an organ is used continuously, it will tend to become highly developed, but if it is not properly used, it will be degenerated.
4. Modifications, acquired by the organisms during their life time are inherited to the offspring.

The Lamarckian doctrine can be described in detail as given below :

1. *Role of environmental factors* : Lamarck believed that environment plays an important role in influencing the organisms and ultimately, structural changes are brought about. Thus, the living organisms react to external conditions and become modified in response to changes in the environment.
2. *Effects of needs* : According to Lamarck change of habits initiates the formation of new organ, also, some

modifications may be seen in the existing organ or bodily structure.

3. *Use and disuse* : Lamarck thought that if the organs are used continuously and constantly, they become well developed and enlarged in size also. But, permanent disuse of these organs makes them reduced and weakened and finally they lead to degeneration. After many generations they may be completely lost. For example Lamarck cited the case of Giraffe. In this African creature, the neck of the ancestor was short because they lived in grassy plain. Their fore–limbs and hind limbs were of equal size. Later on, they had to depend on the leaves of the trees for food. Naturally, elongation of neck and forelimbs was gradually achieved. Thus, in the present day Giraffe we find long neck which is the result of continuous stretching of neck. Of course, this character was acquired after many generations. Similarly, there was elongation of the forelimbs also.

4. *Inheritance of acquired characters* : The changes which take place in the body of the animals during their life time due to direct or indirect environmental effects or *use* and *disuse*, are known as *acquired characters*. These characters are preserved by the animals and these are inherited to the subsequent generations. Transmission of these characters to the offsprings become more and more pronounced if they are exposed to similar stress of the environment like their parents or ancestors. Such cumulative effects may result in the appearance of new species.

Examples based on direct observations in nature

1. The ancestors of modern horses lived in the dense and luxuriant jungles. They used to feed on succulent herbage. That time the rainfall was greater and they found shelter behind the bushes against the enemies. When the climatic conditions changed the two types of adaptations were seen (a) perfection of food getting mechanism; (b) perfection of the mechanism of speed.

   The important adaptations undergone by the horses were—

   (i) Grinding area of the teeth was increased by the conversion of premolars into molars.

(ii) Lophodont and Solenodont appeared

(iii) There was development of cement for strengthening the teeth.

(iv) Lengthening of the head and limbs was achieved.

(v) Change from plantigrade to unguligrade (locomotion)

(vi) Reduction of digits from 5 to 1

(vii) Perfection of hoof

(viii) General size of the limbs increased and the neck elongated

(ix) Large brain with high intelligence

2. The neck of Giraffe was elongated because of the continuous stretching of neck to reach the leaves of the tall trees to make them available as food. It was of course done for several generations, and only then the long neck were achieved in the modern Giraffe.
3. Perching birds have developed sharp and curved digits.
4. Water birds developed webbed feet for swimming.
5. Eyes are reduced and degenerated in the cave dwelling forms.
6. Pinnae are reduced in man but they are well developed in dog and rabbit to collect sound waves.
7. The limbs are absent in snakes because they were not suited for crawling and burrowing. Rudiments of limbs are still present in Python.

Criticism of the Theory

Cuvier and Weismann were the great critics of Lamarck's doctrine (Lamarckism). Weismann performed *mutilation experiment* on mice. He went on cutting the tails of mice for many generations but he never found a tailless mouse or mice with reduced tails. Thus he concluded that the acquired characters are not inherited. For transmission of characters, only germ cells are responsible, as he stated. His germplasm theory had become popular by that time.

Chinese women are using iron shoes for many generations to keep their feet short but the young ones at birth have always normal size of feet. So, this tradition also goes against the *inheritance of acquired characters.*

Similarly, boring of ears and nostrils in Indian women is in practice for centuries, but these characters are never inherited and the offspring have no holes in ears and nostrils at birth.

A blacksmith has powerful muscles (*Bicep* and *tricep*) in the arms because of constant and regular use of the arms, but this character is not inherited. This is an acquired character.

Loeb, with the help of chemical stimuli could show artificial Parthenogenesis in Sea–urchin's egg. Similar eggs were produced in brine shrimp (*Artemia salina*), but in no generation the acquired characters were seen.

Pavlov tried to show the inheritance of learning. Mice were trained by him for this purpose. Mice learnt to come for food on hearing a bell. This training became necessary for all generation. This acquired character was not inherited in any of the generations.

Castle and Phillips performed transplantation experiments on the guinea pigs. The ovary of a black female guinea pig was removed and was transplanted into the body of a white female guinea pig. Now, the white female (with the ovary of black guinea pig) was mated with a white male. It was found that all the offspring produced from this pair were black in colour. The result of this experiment goes against the idea of inheritance of acquired character. The environment has no effect on heredity.

In addition to the above facts, the development of new organ or structure does not depend on mere desire. It is not true that a new organ develops when the organism need it. The environment can cause temporary change only, which can not be inherited.

Lamarck's theory was finally rejected mainly on the ground that acquired characters produce variations which are never inherited. There are several cases of non–hereditary nature of acquired characters.

Neo – Lamarckism

It is a fact that the Lamarckian concept of "*inheritance of acquired characters*" was rejected for sometimes, but, later on, the importance of this doctrine was realised. Some evidences were discovered which favoured this view. The modified form of Lamarck's theory was known as Neo–Lamarckism. Several supporters accepted that the acquired characters are inherited to some extent. Some

Neo–Lamarckians were Cope, Dali, McDougall, Packard, French, Giard, Haeckel, Herbert Spencer, Gadow, Eimer and others. Several evidences were put forward which are as follows —

1. *Bonner* carried out transplantation experiment in all sorts of environmental and concluded that variations produced were inherited to the subsequent generations.
2. *Sumner* reared white mice at high temperature (20°-30°C) and it was found that the mice developed longer body, longer hindlimbs and a long tail. These abnormalities were transmitted to the offsprings.
3. *Brown Sequard* found that several acquired diseases are transmitted from one generation to another. *Exophthalmia* is caused by injuries in the restiform body of the brain and this acquired disease is inherited to subsequent generations.
4. *McDougall* trained white rats to follow certain "escape routes" from a water tank and this experiment was done for 45-50 generations of these rats. He found that there was a progressive decrease in the number of errors made in learning the problem generation after generation.
5. *Kammerer* performed his experiments on *Proteus anguinus*, a cave dwelling amphibian which possessed vestigial and colourless eyes. *Proteus* was brought into day light and it was found that after sometime the development of eye took place and its colour became brown and black. These acquired characters of the eye were inherited to the offsprings.
6. *Tower* found that when potato beetles were subjected to extreme conditions of temperature and moisture, marked changes in the offsprings were observed. Tower found that these changes were inherited to succeeding generation.
7. X–rays produce changes in the chromosome which lead to changes in the offsprings making them different from their parents.

Conclusion

Finally, the Lamarckian doctrine became somewhat acceptable as it was based on the idea of adaptation and direct and casual relationship between structure, function and environment. Forces of evolution must integrate these aspects (mentioned above). "*Mneme*" theory of *Semon* illustrates that every condition of life or functional activity of organism leaves permanent record of itself in the form of "engramme". If these conditions are continued for a long period, these *engrammes* affect heredity and become heritable. Neo–Lamarckians are still looking forward with confidence to the eventual discovery by which it may be confirmed that somatic changes may induce corresponding changes in the germ cells.

Chapter 13

Variations, Recapitulation Theory, Genetic Equilibrium and Hardy Weinberg Law of Equilibrium

Variations

It is a matter of common observation that two individuals of the same species can not be exactly similar. Even the twins are not exactly alike. There must be some differences, which are known as variations. Some variations are at least very important for the process of evolution. *Genetic variations* are the *heritable variations* which have been described later on. Variations can be morphological as well as physiological.

Kinds of variations : There are three sets of variations which have been described below :

1. *Meristic* and *Substantive variations*
2. *Continuous* and *discontinuous variations*
3. *Somatic* (somatogenic) and *Germinal* (blastogenic) *variations*

Meristic and Substantive Variations :

The *meristic variations* are numerical variations. These are variations in the number of parts of an organism e.g. presence of 6 digits in man instead of normal 5. Similarly there are sometimes 6 arms in starfish, though, normally there are 5 arms. The *substantive variations* include changes in structure, shape and size of the organisms. For example, some family members become very tall when others are dwarf. Variations are also called *Orthogenetic* when they occur in an adaptive direction.

Continuous and Discontinuous Variations

The *continuous variations* are very minute and constant variations. It is a slow and gradual variations fluctuating on either side of the average and hence sometimes called *minus* and *plus* variations. Darwin considered these variations significant in the origin of new species. Darwin's concept was that the progressive accumulation of continuous variations contributes in the formation of new species. These variations are acted upon by Natural Selection. There is very little degree of change between successive generations.

The *discontinuous variations* are large sudden and rare. The deviation is to a great extent providing a new appearance. These are also known as *Saltations* or mutations. These variations are heritable and can be transmitted to the subsequent generations. There is no definite periodicity of such variations. There is mutation of form, colour and size e.g. *Short–legged ram* of Massachusetts detected in 1791.

Somatic and Blastogenic Variations

Somatogenic variations are also called acquired variations. These variations are the result of direct influence of the environment on the body of the organisms. These affects the somatic cells of the body. These variations are not heritable and they can not be transmitted to the offspring. Effects of *use* and *disuse* are considered under this head. By constant use of the arms the *bicep* and *tricep* muscles are well developed in the blacksmiths. But, this character is not seen in the progeny. Lamarck laid stress on the inheritance of acquired characters but later on, this idea was disapproved by Weismann.

Blastogenic variations occur in the germ plasm of the organisms and they are passed on to the offspring. Such variations appear because of recombination of genes or due to mutations. Mutation may occur at any time in the life of an organism.

The sources of variations are Environment, Endocrine Glands, Mutation and Recombinations of genes. Among these sources, the mutation is the most important factor which has been described in the "Mutations" chapter. Both gene mutations and chromosomal mutations have been described in detail there.

Genetic Equilibrium and Hardy Weinberg Law of Equilibrium

The fundamental idea in population genetics and modern

evolutionary theory was proposed by G.H.Hardy and W.Weinberg simultaneously in 1908. It was popularly known as *Hardy–Weinberg Law.*

Hardy–Weinberg's law describes the situation in which there is no evolution, and it provides a theoretical baseline for measuring evolutionary change. The equilibrium tendency avoids too rapid changes and conserves gains made in the past. Distribution of genes and genotype frequencies have been well explained here.

The Law states that "*the relative frequencies of various kinds of genes in a large and randomly mating sexual panmictic population tend to remain constant from generation to generation in the absence of mutation, selection and gene flow*".

This law explains a theoretical situation when there is no evolutionary change in the population. In absence of evolutionary forces — (i) when the population is large (ii) individuals exercise random mating, (iii) each parent produces almost equal number of gamets, (iv) the gamets produced by the mates combine at random, (v) and the gene frequency remaining constant then the genetic equilibrium of the genes in question is maintained and the variability present in the population is well preserved.

Let us presume that there is a panmictic population with gene A and a on one locus. The frequency of gamets with gene A will be the same as the frequency of gene A. In the same way, the frequency of gamets with a, will be equal to the frequency of the gene a.

Suppose that the numerical proportion of different genes in this population is as given below :

AA	Aa	aa
36%	48%	16%

Thus AA individuals will contribute 36% of the gamets of the populations possessing gene A. In the same manner aa individuals will be producing 16% of the gamets. But, the Aa individuals will form the gamets with both gene A and with gene a (in both cases 24%) as given below :

Parents	*Gamets*	*Parents*	*Gamets*
36%AA	36%A	16%aa	16%a
48%Aa	24%A	48%Aa	24%a
Total	60%A	—	40%a

Because of random union of gamets, the total number of different genotypes will be as stated below :

Sperm	*Ova*	*Gene frequency*	*Offspring*
A	A	60 x 60	36%AA
A	a	60 x 40	24%Aa
a	A	40 x 60	24%Aa
a	a	40 x 40	16%aa

Suppose that the frequency of a gene A is p, and that of gene a is q. If the random union of gamets continues with A and a at the state of equilibrium, the population will contain following frequencies of A and a genes generation after generation —

AA + 2 Aa = aa genotype

$p^2 + 2pq + q^2$ gene frequency.

These results are explained on the basis of *probability theory*. In large population, the probability of receiving the gene A from both parents will be $p \times p = p^2$, and for gene a it will be $q \times q = q^2$. The probability of heterozygote will be $pq + pq = 2pq$.

So, the relationship between the gene frequency and genotype frequency is expressed as –

$$p^2 + 2pq + q^2 = 1$$

$$(p + q)^2 = 1$$

The binomial expression is known as *Hardy–Weinberg formula*. In a large and randomly mating population the gene frequencies and the genotype frequencies remain almost constant. In the small populations, gene frequencies are unpredictable.

In the absence of all evolutionary forces, the gene and genotype frequencies of each allele in a population at an equilibrium state generation after generation. Reproduction is equally successful in all the genotypes. Mating in a large population is always random. Hardy–Weinberg's law also states that particular alleles are neither differentially added to, nor differentially subtracted from a population.

Recapitulation Theory

The theory proposed by Ernst Haeckel (1811) is popularly known as *Recapitulation theory* or *Biogenetic Law* of Haeckel. This theory is illustrated as "ontogeny recapitulates phylogeny"

Ontogeny means the development of the individual, and phylogeny means the history of the race. So, the developmental history of the individual repeats the history of the race or species. The embryonic stages of the individual resembles the adult stages of the ancestor. For example, the zygote corresponds to a single celled Protozoan ancestor, the blastula corresponds to colonial flagellates and gastrula represents two layered animals like coelenterates e.g. Hydra. The establishment of three germ layers corresponds to flatworms. The appearance of chordata characters like tubular dorsal nerve cord, notochord and gill slits provides points of resemblances with the primitive chordates and so on.

It becomes clear from the sequence of action during the course of development that there are more similarities in the early ontogenic development of various groups of animals in comparison to later stages. Various stages (either earlier or later) of ontogenic development can be affected by mutations.

Though the Biogenetic Law was based on some genuine facts, however, some objections were raised against this theory. Later on, it was shown that the resemblances are mostly between the embryonic states of related animals. Chief opponents of this Recapitulation theory was Garstang and DeBeer who claimed that similarity in ontogeny between the animals is simply an indication of their affinity. Embryos show several adaptations and modifications in their developmental process. There is no question of similarities with the ancestral adults.

Von Baer's Principle

Von Baer's theory proposed earlier than Recapitulation theory of Haeckel has been claimed to be more appropriate. This theory states —

1. General characters appear first during development and special features afterwards.
2. Less general characters also appear, but specific characters of the class appear first, then those of order, then of family and then lastly, the characters of species appear.
3. Embryos of one species differ from the embryos of other species. Thus, the differences are more important than resemblances. There is progressive departure of the embryos of one animal from that of the other.

4. That the embryonic stages of individuals show resemblances with the embryonic stages of ancestors and not the adult ancestors.

Modern Concept of Recapitulation Theory

Swereizoff a Russian Scientist has pointed out the relationship between developmental stages and ancestral adults in detail. It is known as *Phylembryogenesis*. The events have been described under the following heads :

1. *Addition, Prolongation or anaboly*: In this case of process of development follows the ancestral pattern upto the last ancestral embryonic stage and further stages are added after that.
2. *Deviation*: Here, there is deviation in the process of development as compared to the ancestor. The embryonic development of the descendent follows the ancestral pattern only up to certain stages and after that the events get modified.
3. *Archallaxis*: There is deviation in the embryos of the ancestral and descendent species from the very beginning. Recapitulation has not been marked in this case.
4. *Abbreviation*: That, some of the embryonic stages of ancestral development which make their appearance at the end of the process are omitted or deleted during the development of descendent. The process of *Neoteny* is an example of this type, where the larva omits the process of metamorphosis. Development of gonads takes place and the larva behaves like the adult.
5. *Acceleration*: In this type of ontogeny at least some of the middle stages of embryonic development of the ancestor are omitted in the descendent ontogeny.
6. *Coenogenesis*: Here the non–primitive characters make their appearance in the consequence of secondary adaptation of the young to the peculiar condition of the environment. Thus, appearance of specialised or adaptive characters is described as coenogenesis. Mimicry in insects may be cited as an example in this case.
7. *Omission*: Sometimes not only the early embryonic stages of ancestral development are omitted but the latter stages also deviate from the ancestral stages.

8. *Heterochronism*: In this case, sequence of events may be changed in the development of descendent. Some of the organs appear either earlier or later in the descendent embryos as compared to the ancestral ontogeny.

Importance of Recapitulation Theory

Recapitulation theory helps in establishing the relationship of several animals whose adults stage provides no clue of their phylogenetic relationship, as in the case of *Sacculina, Herdmania* etc. The ancestry in certain groups is also ascertained with the help of recapitulation theory which provides an evidence in favour of organic evolution describing the course of development by the ancestors.

Palingenesis

"Truly ancestral characters conserved by heredity are produced in development". For example, the tadpole larva of frog has some characters like gills, fins, lateral line system etc. which are characters of fish, thus indicating that amphibians have evolved from piscine ancestors. After metamorphosis these characters are lost in the adult.

Tachygenesis

"The acceleration of characters which are crowded back further and further into the embryonic life or out of life cycle entirely" — is known as *tachygenesis*. There is condensation of some of the phylogenetic events and the elimination of others in the embryonic development. Suppressed larval stage in some of the eggs of frog is an example of tachygenesis.

Chapter 14
Adaptations

According to Darwin's Natural Selection theory, only most viable and adaptable species survive in nature and others are eliminated. Thus, adjustment of the species to their new surroundings is a very important factor in the formation and conservation of species. Thus, *Adaptation is a morphological, anatomical and physiological modifications in an organism necessary to adjust itself in particular environment.* In the process of adaptation, acquisition of adaptive features through the interaction of evolutionary forces take place. Adaptations are basically of two categories i.e. *Convergent adaptation* and *Divergent adaptation.* In different diverse forms, similar adaptations concerned with locomotion, food getting and offense and defence occur. So, there is convergence at one feature. For example, for leading a volant life, the animals of different groups will have to develop wings. This is a case of *convergent adaptation.* The divergent adaptations are just contrast to this. This is also known as *adaptive radiation.* Here, the animals of the same group or closely related groups exhibit divergence in their morphological features and they look quite different when they are found in different habitats. Thus closely related genera become adapted to different kinds of environment. Adaptive radiation in mammals is an example of divergent evolution. It was called *embranchment* by Lamarck.

Kinds of adaptations: There are four important kinds of adaptations. These are —

1. Structural adaptations
2. Physiological adaptations
3. Protective adaptations
4. Animal association adaptations

Now each adaptation has been described in detail below :

1. Structural adaptations :

Structural adaptations may be *morphological* or *anatomical.* Organs are adapted according to the needs of the body in response to particular environment. According to Mathew, the structural adaptations may be classified as follows —

1. *Cursorial adaptations*
2. *Fossorial and Subterranean adaptations*
3. *Scansorial (arboreal) adaptations*
4. *Volant adaptations*
5. *Aquatic adaptations*
6. *Cave adaptations*
7. *Deep–sea adaptations*
8. *Desert adaptations*

1. *Cursorial or terrestrial adaptations*: These adaptations are meant for running swiftly on land. So, these are also known as speed adaptations. Various examples of swift runners are — snakes, lizards, extinct dinosaurs, emus, ostriches, rhea, kangaroo, horse, dogs, cats, leopard etc. The characteristic features are —

    (i) *Streamlined body* : The body is spindle-shaped so as to offer least resistance while passing through the air, e.g. Horse.

    (ii) *Less development of forelimbs* : In most of the cases the forelimbs are less developed but the hindlimbs are modified as they contribute in fast locomotion.

    (iii) *Digitigrade* foot : The animals run fastly on digits. Wrist bones are raised from the ground. Sometimes they possess muscular pads, as in dogs. They are generally 4 toed.

    (iv) *Unguligrade locomotion* : Modified nail or hoof is present for swift walking. The distal bones are depressed or flattened. They reach highest perfection in horses. In some desert animals (e.g. camel) the cushion like pad is present. They have become secondarily digitigrade. They have 1 to 4 toes.

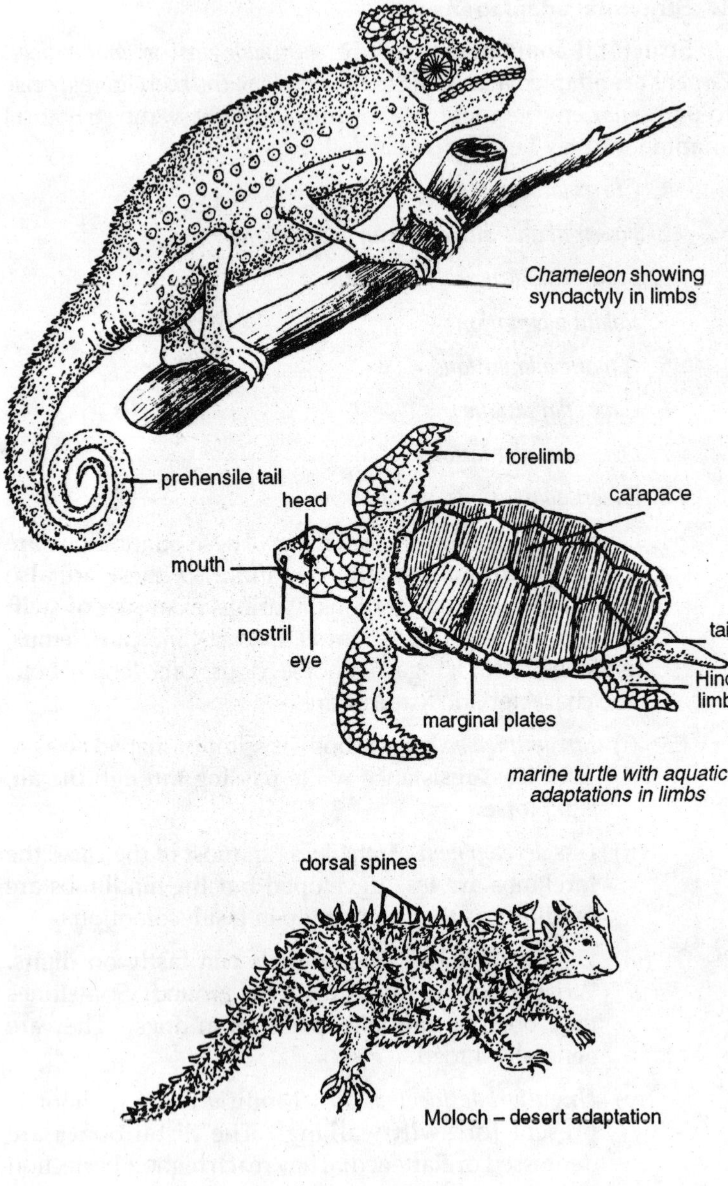

Fig.42 : Some reptiles with different adaptations

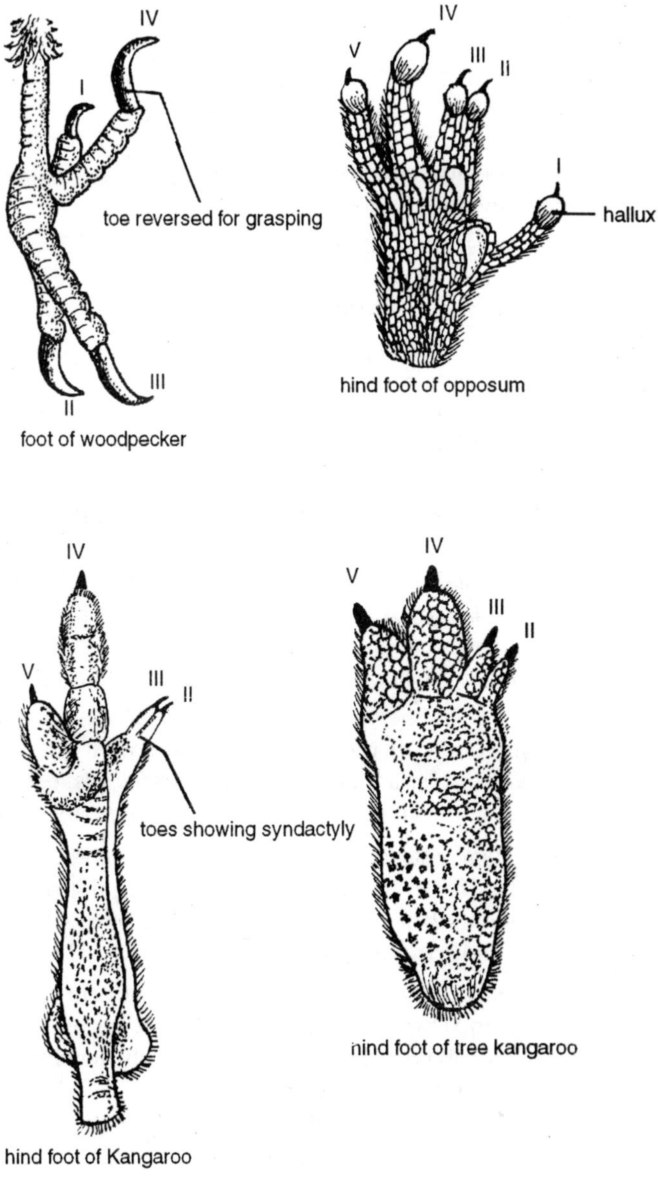

Fig.43 : Hind feet of a bird and marsupials

(v) *Loss of digits* : In plantigrade forms the entire palm or sole rests on the ground as in Bears and some Primates. They are normally 5–toed. Speed adaptation influences the digital reduction.

(vi) *Reduction of Ulna and Fibula* : Ulna and fibula bones respectively in the forelimb and hindlimb are considerably reduced (e.g. in Horse). Clavicles in several animals tend to reduce.

(vii) *Lengthening of distalia of limbs* : The elements of the limbs are elongated for better speed. The muscles are thick and short.

(viii) *Movements of limbs in one plane* : The limbs can be moved only in one plane at the joints (except ball and socket joint of hip and shoulder). Interference between forelimbs is avoided during running. For example, in case of dog the hind limbs come forwards and outside of the forelimbs when the progression is speedy.

(ix) *Bipedal changes* : In the bipedal forms like man, birds and kangaroo, there is reduction in the size of the forelimbs in comparison to hind limbs. In the semi-erect bipedal forms the tail is a prehensile organ.

2. *Fossorial and Subterranean adaptations* : These are changes in the body of the burrowing animals for example — *Gryllotalpa*, spiny anteater, Hog–nosed, snake, *Condylura* (Common mole). The important changes are —

(i) *Spindle–shaped body* : The body is spindle–shaped or fusiform so as to offer least resistance to the subterranean passage. In certain forms the body is cylindrical (e.g. snakes, caecilians).

(ii) *Tapering head* : Tapering of head anteriorly is an important feature which forms a snout for burrowings. Jaw muscles are poorly developed. Neck is either absent or very short.

(iii) *Reduction of tail* : The tail is vestigial (as in moles) and it may serve as a tactile organ. The tail is short in hedgehogs.

(iv) *Disappearance of Pinnae* : The external ears disappear (e.g. Monotremes) as it would become an obstruction in the digging process and living in the burrows.

(v) *Loss of eyes* : The eyes are of no use in the dark habitat and as such they are vestigial (e.g. in Talpa). The eyes may be covered with skin (e.g. Golden mole). In several animals it is very much reduced in size.

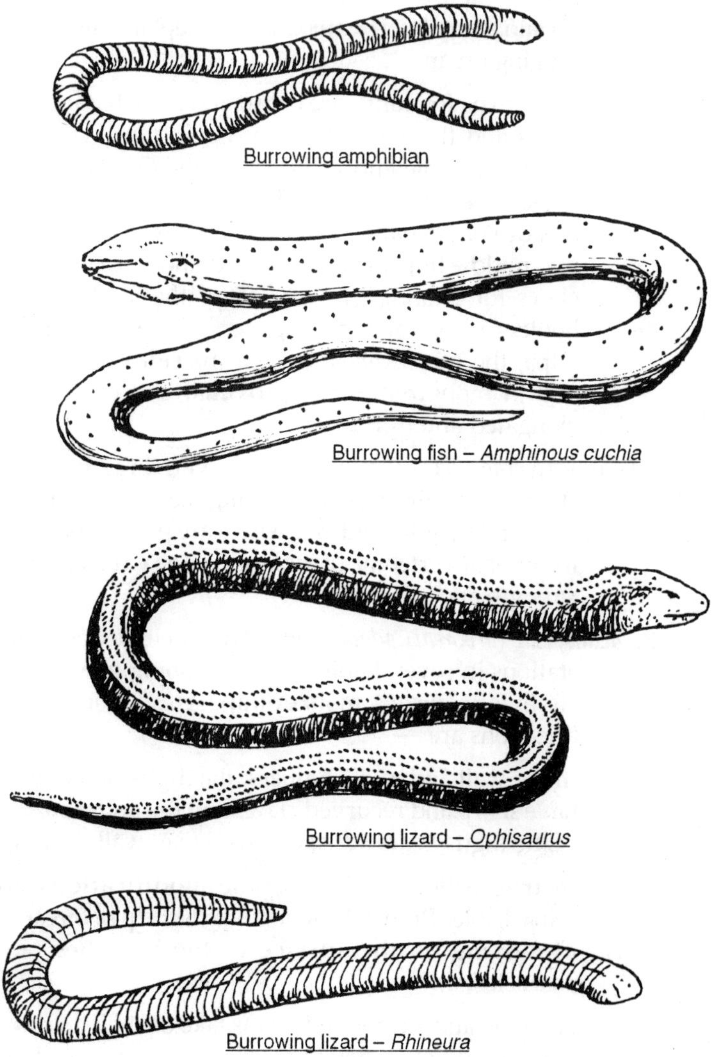

Fig.44 : Animals with fossorial adaptations

(vi) *Presence of digging apparatus* : Various structural changes take place because of digging habit. The snout is well developed and turned upward at the tip (e.g. Swine). Sometimes, prenasal bone develops at the tip of the nose cartilage (e.g. Talpa). Incisor and canine (for example in Swine) teeth are fossorial organs in certain animals. The tusks of the elephant also act as digging organ.

(vii) *Limbs* : The forelimbs are broad and stout having long claws and they are useful in digging (e.g. common mole). In mole and echidna a bone *Os falciforme* is present in the palm in addition to 5 fingers and the breadth of the palm is increased. In the golden mole two middle fingers are elongated and have powerful claws for digging. In *Gryllotalpa* the forelegs have hooks for burrowing purpose. In the hindlimb of talpa, there is *sesamoid bone* at the sides of the tibia and in monotremes the proximal end of fibula is elongated anteriorly.

(viii) *Vertebrae* : The fusion of sacral vertebrae provides strength and firmness in pushing the animal through the burrow. Intercentral ossicles are present between the lumbar vertebrae of mole and talpa which provide strength and rigidity.

3. *Scansorial (arboreal) adaptation* : This includes climbing adaptations in several animals i.e. climbing perch, hyla, lizards, flying squirrels, monkeys, apes etc. The important modifications are —

(i) In the wall and rock climbers, the digits of the limbs have sharp and recurved claws and the under surface of the digits bears adhesive discs (e.g. wall lizards).

(ii) In true arboreal forms some modifications are remarkable. Branch runners take the help of forearms in moving on the surface of the branches (e.g. chameleons).

(iii) Hanging and suspended forms take the help of claws of the hind limb when they hang (e.g. Bats), or claws of the forelimbs (e.g. Sloths). Apes are seen swinging with great speed and accuracy from limb to limb and tree to tree.

(iv) The dorso–lumbar vertebrae are elongated in tree slothes and there is increase in number (25–27). This is an arboreal adaptation.

(v) The girdles become tough and strong to support the weight of the body. Clavicles and scapulae are well

Fig.45 : A tarsier, a lemur and a new world ape showing arboreal habit

developed. The pelvic girdle has broad ilium to support the viscera (e.g. sloths).

(vi) Humerus becomes elongated is suspended and brachiating forms e.g. in sloths and gibbons.

(vii) In some arboreal forms the limbs are of grasping type with opposable digits but in others they are non–prehensile. In chameleon the digits are arranged in two opposable bundles of 2 and 3 digits in the limbs. These help in holding the branches of the trees. It exhibits *Syndactylous* condition. In parrots, wood–packers etc. the outermost 4th toe rotates backward in such a way that this one and hallux become opposed second and third digits. Thus, two toes are directed forward and two backward. It shows *Zygodactylous* condition.

(viii) In the arboreal forms, there is reduction of digits. In *Koala*, second and third toes are *Syndactylous*. In tree sloths only two digits are present in the forelimbs and three in the hind limbs. The second digit is absent in some lemurs.

(ix) The tail may be prehensile as in Chameleon and some monkeys. It may be non–prehensile also. Sometimes, ecodermal spines are present beneath the tail which prevent the animals from falling down (e.g. flying squirrel).

(x) Different types of accessory climbing organs are developed. In *Lemur catta* there is a climbing organ of hardened skin in the forearm.

4. *Volant (aerial) adaptations* : It is also known as aerial adaptations. The animals are capable of flying in the air. Some animals can fly in the air for a longer time and some can glide in the air for a short period. The former is known as *active flight* and the latter is called —*passive fight*. Active flight can be seen in insect, birds and some mammals (e.g. Bats). Modification of forelimbs into wings, presence or feathers, pneumatic nature of bones, fusion of bones, development of powerful flight muscles etc. are some of the flight adaptations found among birds. There are modifications practically in every system of the body of

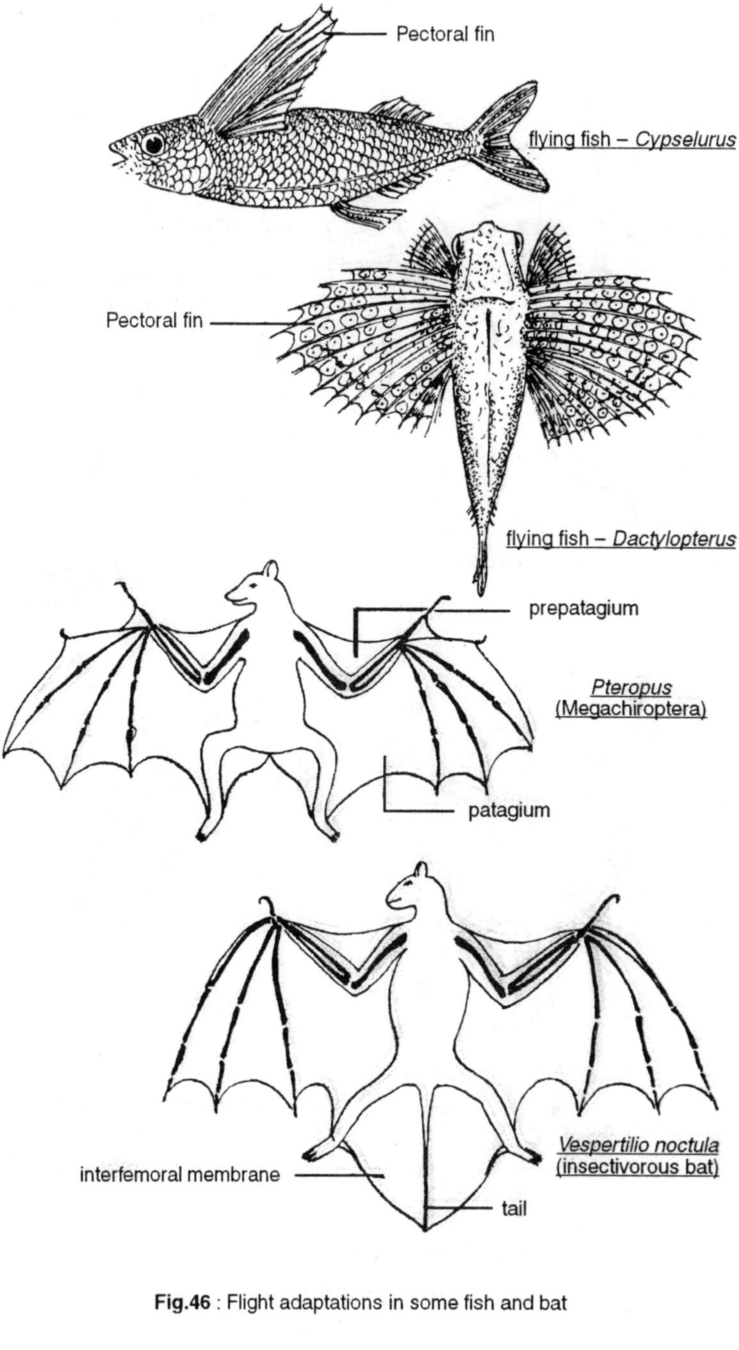

Fig.46 : Flight adaptations in some fish and bat

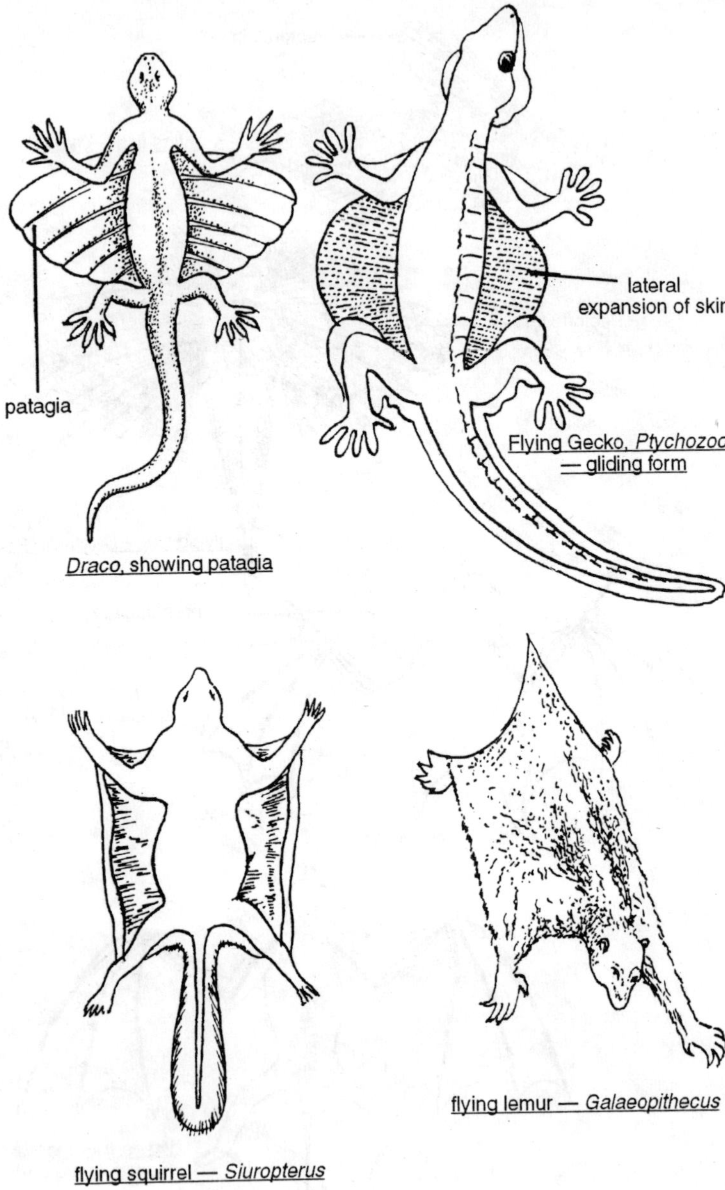

Draco, showing patagia

Flying Gecko, *Ptychozoon* — gliding form

flying squirrel — *Siuropterus*

flying lemur — *Galaeopithecus*

Fig.47 : Some animals with flight adaptations

the birds. For detailed description on *"flight adaptations in birds"*, (Class — AVES chapter) of this book can be seen. Active flight is also seen among the bats. These are the only mammals which exhibit true fight. For this purpose the wings are present which are extensions of skin on either side of the body. There is development of pectoral muscles also. Other modifications in the bodily organization of bats related to flight have been described in *'Chiroptera'* (Class – 'Mammalia" chapter) of this book.

Active fight is also exhibited by insects which are equipped with wings. The wings in insects are thin folds of cuticle which are supplied with veins and very fine nerves. They are operated by muscles.

Passive flight (Gliding) is seen among some fishes, amphibians, reptiles and even some mammals. In *exocoetus,* a flying teleosts fish, the pectoral fins are modified to form a wing like structure with the help of which, the fish can take short flight in the sea. Among amphibians, *Rhacophorus,* can glide from tree to tree with the help of webbed digits of the limbs. Also, they have adhesive pads at the ends of the digits of the limbs for adhesion to trees. Among the reptiles, the flying lizard, *Draco volans* can glide with the help of *Patagium,* the fold of the skin beneath the fore limbs and hind limbs. Some mammals can also glide from one tree to another, with the help of patagium. The flying lemur, *Galeopithecus volans* is an example of soaring mammal.

5. *Aquatic adaptation :* The structural modifications which enable the animals to live in water are known as aquatic adaptations. Water is the home for a large number of animals practically from every group of the animal kingdom. For details *"Aquatic adaptations in Mammals"* and *"Cetacea"* (Class — Mammalia chapter) of this book can be seen. Presence of fins, air bladder and lateral line system are supposed to be important aquatic adaptations in fishes. The lateral line system contains the *neuromast organs* which act as *rheoreceptors* meant for the purpose of ecolocation of objects in the aquatic medium.

6. *Cave adaptations :* Caves are mostly the abandoned channels through which rivers were flowing in the past.

They are present in the limestone region. Lateral and subterranean channels are also formed due to the lateral movement of water. These abandoned channels are called *caverns*.

There is absence of light in the cave, except near the mouth of the cave. Nearly, uniform temperature is there, but, seasonal fluctuation of temperature may take place. There is scarcity of food and high humidity in the cave. Also, there is lack of competition and scarcity of natural enemies.

White–footed mouse (*Peromyscus leucopus*), oilbird (*Steatornis caripensis*), *Siren*, *Proteus anguinus*, *Bathynella* (a crustacean), *Amblyopsis spelaeus* (a fish), *Typhlotriton* and *Ichthyophis* (both amphibians) are the important common fauna of a cave. The important adaptations of the cave dwellers are :

(i) *Loss of Pigmentation* : Because of absence of light the animals are mostly colourless. Sometimes pale with white colour is seen. *Proteus* develops dark pigments when it is brought into light.

(ii) *Slender body and appendages reduced* : The cave dwellers have a slender body and the appendages are attenuated. In *Bathynella* the antennules and antennae are slender. In urodels, the body is slender.

(iii) *Weak exoskeleton* : The animals of the phylum–Arthropoda as a whole, the exoskeleton is weak and fragile. The cave fishes are mostly without scales. The snails have very thin shells. The insects are also very weak and thin.

(iv) *Absence of eye* : There is absence of visual organs in the animals living in caves. Eyes are either reduced, degenerated or totally absent. It may be non–functional also (e.g. in *Typhlotriton*).

(v) *Tactile organs* : Tactile organs are well developed to compensate the absence of visual organs. In myriapods and arachnids the legs and antennae have tactile hairs. Also, these are very long. The head of *Amblyophis* bears taste organs in the form of buds. Chemoreceptors are present in *Asettus* (— an isopod) as sense bulbs.

7. *Deep–Sea adaptations* : Deep–sea (from 6000–10000 metres deep) has unique ecological features. The sun light does not penetrate to a great depth in the sea. So, there is ever ending darkness. The temperature is low and there is no seasonal fluctuations. There is excessive pressure exerted by the water column in the deep sea so the normal life is

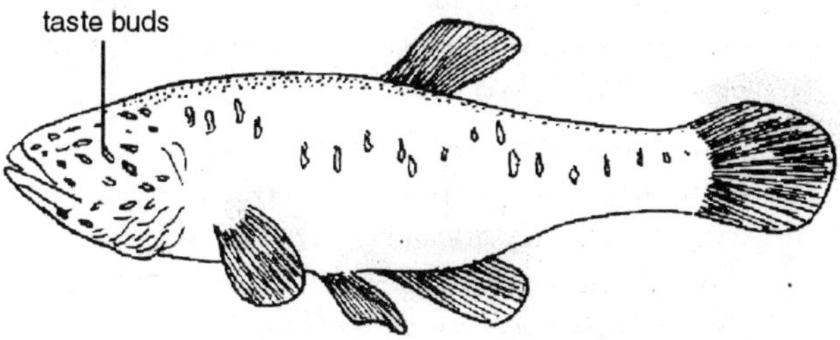

Cave fish — *Amblyopsis*

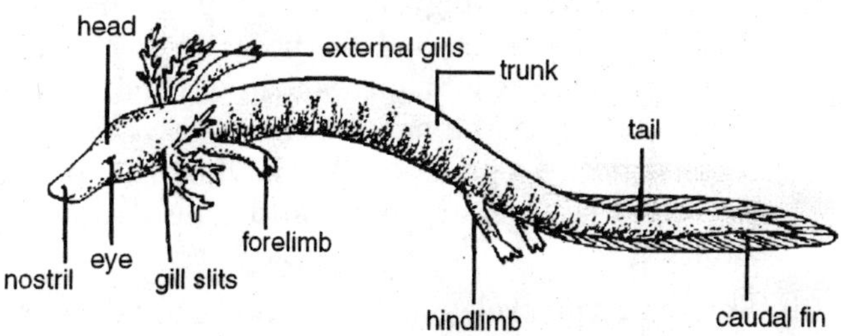

Proteus (an amphibian)

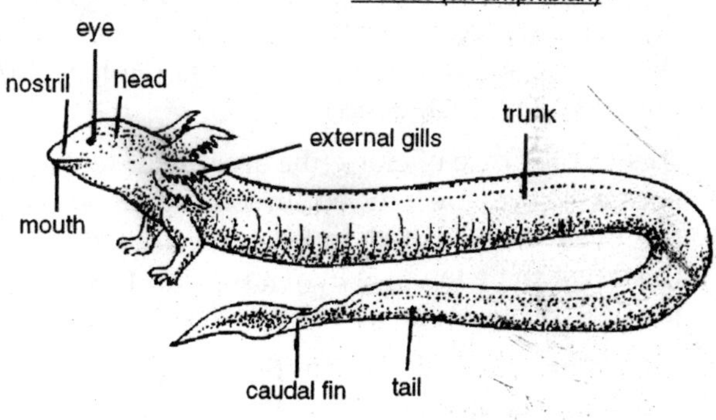

siren

Fig.48 : Animals with cave adaptations

very much affected. There is no water current and there is perfect calm in the deep-sea region. Sediments of this region consist of skeletons of various pelagic forms.

Some radiolarians (e.g. *Acantheria*), sponges (*Hyalonema, Euplectella* etc.), Coelenterates (Sea–pen, Sea–anemones etc.), Annelides (e.g. *Arenicola, chaetopterus*), Arthropods (e.g. King crab, Lobsters and Shrimps), Molluscs (e.g. Chitons, gastropods), Echinoderms (e.g. Starfishes, Sea–urchins and several crinoids) and Fishes (e.g. *Chimaera* and *Harriotta–Holocephali*), *Photostomias* and flat fishes are the important fauna of the deep sea region. The important deep–sea adaptations are :

(i) *Small weak and fragile body* : The sea–animals are generally smaller in size, of course, there are some exceptions (e.g. *chimaera*). Though the body is weak but the jaws are larger with cruel armaments as in silver shark (*Gastrostomus*). Legs are long as in crabs, and the stalk is long and delicate as in crinoids. The glass sponges are fragile. Some deep–sea fishes are laterally compressed and the body becomes band–like or ribbon–like in shape. Sometimes both the eyes come to lie on the same side (e.g. *Synaptura*). *Saccopharynx* is a predatory fish having wide mouth, powerful jaws, sharp teeth and long distensible stomach.

(ii) *Skeleton* : The skeleton is soft and non–calcareous. Calcium can not be synthesized at such a low temperature. In some protozoans and corals the exoskeleton is made up of Silica. Molluscs have fragile shells. The endoskeleton is *Chimaera* (Holocephi) is uncalcified (cartilaginous).

(iii) *Eyes* : In the case of crabs, the eyes are absent and in pecten the eyes are reduced. In some fishes, however, the eyes are adapted to perceive dim light. In *Gigantura* the telescopic eyes are present. It can see the nearby objects as well as the distant objects.

(iv) *Bioluminescence* : Production of light is the characteristic feature of the deep–sea animals. It is very common among crustaceans, coelenterates, some starfishes and fishes etc. Light producing organs may be present either on the whole body or they may be localised in particular region (head, sides of belly, oesophagus etc.) of the body. There are many uses of bioluminescence. The light emitted attracts the prey, so it is an enticement. Luminescence can confuse the predators. It can illuminate surroundings also.

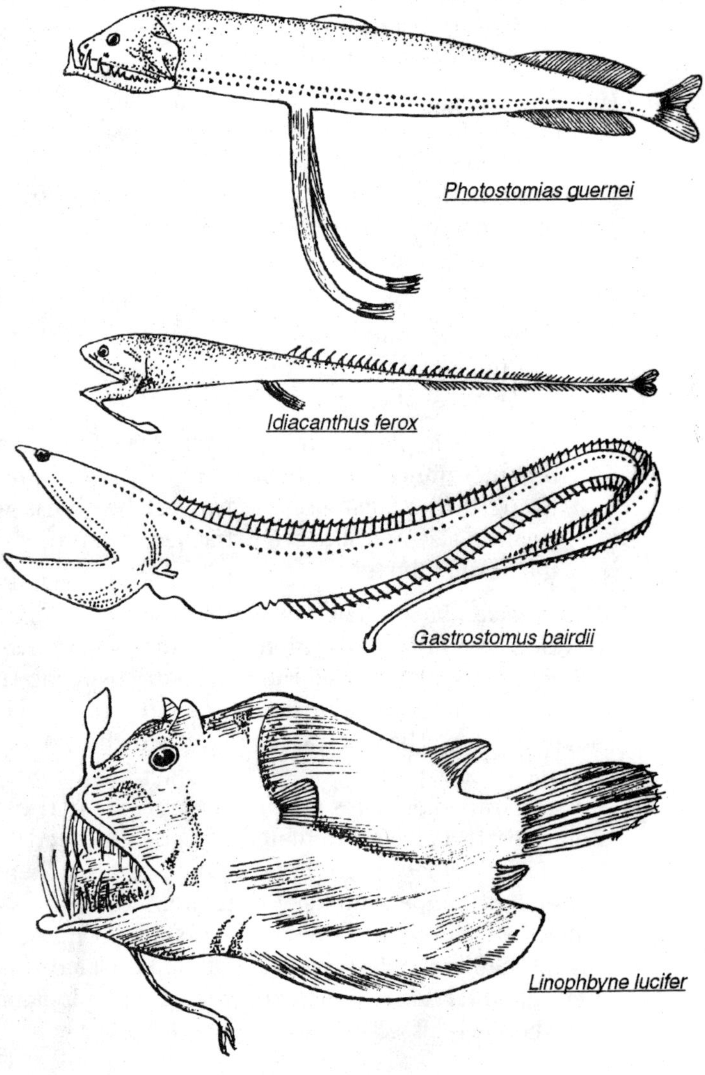

Fig.49 : Deep sea fishes

Sometimes it serves as a warning signal. It is also recognizable character and particular light is distinguished by the members belonging to that species. Lastly the ventral luminescence obliterates the object as in deep–sea *Isospondyli*. Some fishes like Angler fish, *Ipnops*, *Linophryne*, Lantern fish, stomatoids etc. have luminescent organs.

(v) *Lateral–line system* : Lateral–line system is well developed in the deep–sea fishes. It compensates the deficiency of the visual organ.

(vi) *Sensory organs* : In the deep–sea crustaceans there is development of long feelers or antennae. These are even ten times greater than the length of the body. In some fishes the fin rays are produced into long filaments (e.g. *Bathypterois*). These organs are very sensitive to touch.

(vii) *Parental care* : Many deep–sea fishes take care of the fertilized eggs and the young ones. In several cases, enormous number of young ones are produced for the survival of the race.

(viii) *Sexual dimorphism* : In the darkness of the sea it becomes difficult to search the mate when required. In angler fish (*Photocorynus spiniceps*) the male is very small in size and it is seen attached to a process on the head of the female.

8. *Desert adaptations* : The animals living in the deserts develop several modifications in response to *moisture getting* and its *conservation*. The animals are also concerned with self–defence against physical and organic environments. There is scarcity of water in the deserts. It is purely a sandy area. Sometimes, water can not be seen even in traces for miles together except for short oasis. Some deserts fail to get rainfall for several years and there are deserts where there is no rainfall at all. There is extreme of temperature, say, even upto 134°F in Death valley of California. The scorching sunlight makes the life difficult in the deserts. The frequent dust storms are the regular feature of the desert. The various adaptations have been discussed below.

(i) *Moisture getting* : The deep ground water may be available at certain places. Normally, the surface sand is dry and burning. In Australia a few springs are there in the deserts, however, to get sufficient moisture is the prime need of the desert forms. Some animals absorb dew drops alongwith the plant food. The drinking water need is fulfilled by eating succulent plants by the wood rat (*Neotoma*). The desert lizard, *Moloch* has hygroscopic skin which has got the capacity of absorbing water like blotting paper. Some animals eat plant seeds to quench their thirst e.g. kangaroo rat (*Dipodomys*). For the carnivorous animals, their prey is the only source of water. In ants and Kangaroo rats water is obtained from fat, carbohydrates and protein during the metabolic activities.

(ii) *Water conservation* : Desert animals avoid loss of water by evaporation through the body surface. The number of sweat gland is reduced or they may be absent totally. Scales and spines are present on the exposed surfaces of the body which prevent loss of excess water. (e.g. in case of *Moloch* and *Phrynosoma*). Some animals burrow in the sand during day time just to avoid evaporation of water from the body surface (e.g. desert lizard). Several desert animals are nocturnal. Water is stored in the large intestine of *Uromastix* and rumen of the camel.

(iii) *Protection against Scorching Sun* : The animals have protective armour of scales, spines or dermal scutes. Some of them are very swift runners and they manage little contact with the burning sand. Beetles and lizards are well acclimatized to the high temperature of desert. The hind limbs are longer in the desert animals (e.g. Kangaroo rat). The feet are provided with pads as in camel and ostrich. The pads are convenient for movement on sands.

(iv) *Protection against Storms* : Dust storms directly affect the nostrils, ears and eyes of the animals. In camels eye lashes are present in large number and they are quite prominent. Nostrils can be closed like eye lids. Ears have sufficient hairs to protect it. The eyes of typhlops are overhung by the head sheath. In sand swimming lizards the nostrils are turned upwards and they can be easily protected from sand. It is convenient is burrowing also. In *Mabuya* the lower eye lid which is large and transparent can be drawn on the eye to protect it during the storm.

(v) *Self–defence* : Possession of venom is an adaptation for self–defence. Snakes have poison, and fangs associated with it (e.g. rattle snake). Poisonous spiders and Scorpions are present. *Heloderma punctatum*, a poisonous lizard is commonly found in the arid zone of Mexico. Colouration in desert animals is also an adaptation. The body hues are grey, brown, and sandy generally to resemble the general colouring of the landscape which may be deemed to be self–defence feature.

2. Physiological Adaptations

Physiological adaptations are not so obvious as the structural adaptations, but it is a fact that there are also some adaptations in the internal functions of the animals to change conditions of life. In camel water is lost from various tissues but it is not lost from the blood. In the body fluids of Elasmobranchs, there is higher concentration of urea and as such possess higher osmotic pressure. The process of oxidation of fat, carbohydrate and protein is useful for some desert animals in the sense that water is released in these processes. It has been estimated that 1 gr. of fat yields 1.1 gr. of water, 1 gr. of carbohydrate yields 0.6 gr. of water and 1 gr. of Protein yields 0.3 gr. of water. So, maximum amount of water is obtained in the oxidation of fat and minimum in the case of protein oxidation. Balance of chloride ions in the marine fishes by the activities of chloride secreting cells of the gills is an example of physiological adaptation.

3. Protective Adaptations

Protective adaptations are measures of defence against enemies and they also help in getting food. For example, birds have beaks and claws for these purposes. Turtles and armadillo possess an armour of bony plates and hard epidermal scales. Dolphins have recurved teeth to hold fishes which they have to swallow entire. Nature of teeth also depends on the feeding habits

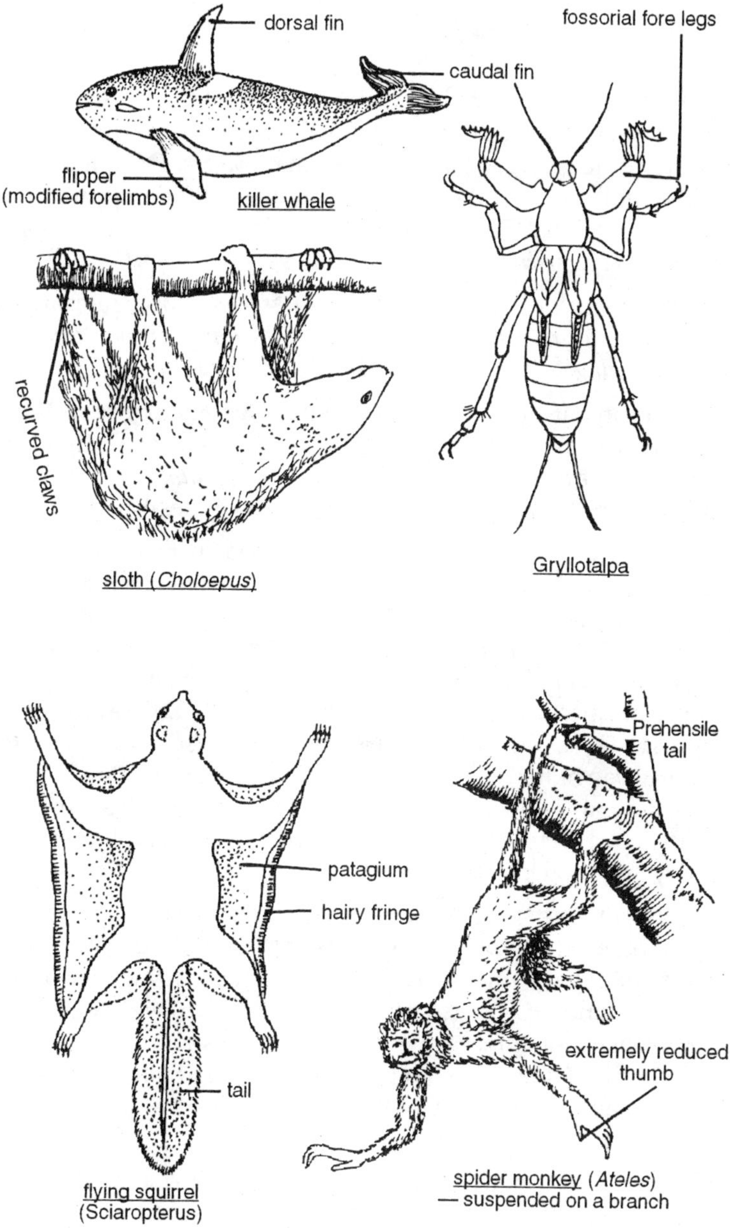

Fig.50 : Adaptations in some animals

of animals and they also act as the organs of offence and defence. Several animals exhibit various contrivances to delude the predators and protect themselves.

Mimicry and Protective Colouration

Mimicry is an act of imitation. It is protective resemblance of an animal to an organism or an inanimate object. Of course, there are several definitions of mimicry, but the definition provided at the International Zoological Congress in Washington in 1963 is very much convincing. It is as follows —

> *"Mimicry is the superficial but close resemblance of one organism to another or to natural objects among which it lives, that secures it concealment, protection or some other advantage".*

The organism which mimics is known as *mimic* or *mimetic* and the one which is imitated or copied is called a *model*. Mimicry is classified into *Protective mimicry, Warning mimicry, Aggressive mimicry* and *Conscious mimicry* (simulation of death).

Protective mimicry is a highly specialised character among the organisms. There are several animals which mimic themselves in form and colour to protect themselves from their enemies. The leaf insect *Phyllium* is a very common example of this type. It is just like a leaf in appearance. The body is flattened, wings and limbs are also leaf like. The colour is green and there are yellow spots. Thus, the enemies are easily confused. *Kallima*, the Himalayan butterfly is another example of Protective mimicry. It is commonly known as deaf–lead butterfly. Its wings are brightly coloured. When it rests with folded wings its underside resembles a dry leaf showing the same colour.

In *aggressive mimicry*, the mimic is an aggressor and preys upon the model. This type of mimicry is very common in carnivorous forms like spiders and fishes. They can not be distinguished from the surrounding objects. They hide themselves and attack their prey when time comes. The animal possesses some lure to attract the prey and it blends with the surroundings. Some animals fall victim and form the prey of the mimic (*Alluring Mimicry*).

In *Warning mimicry* the non–poisonous and harmful animals mimic the poisonous or harmful animals. Some non–poisonous snakes (Elopidae family) resemble cobra and thus, they can deceive the enemies and escape themselves. *Danaus*, the monarch butterfly is distasteful and inedible to birds. It is imitated by edible viceroy

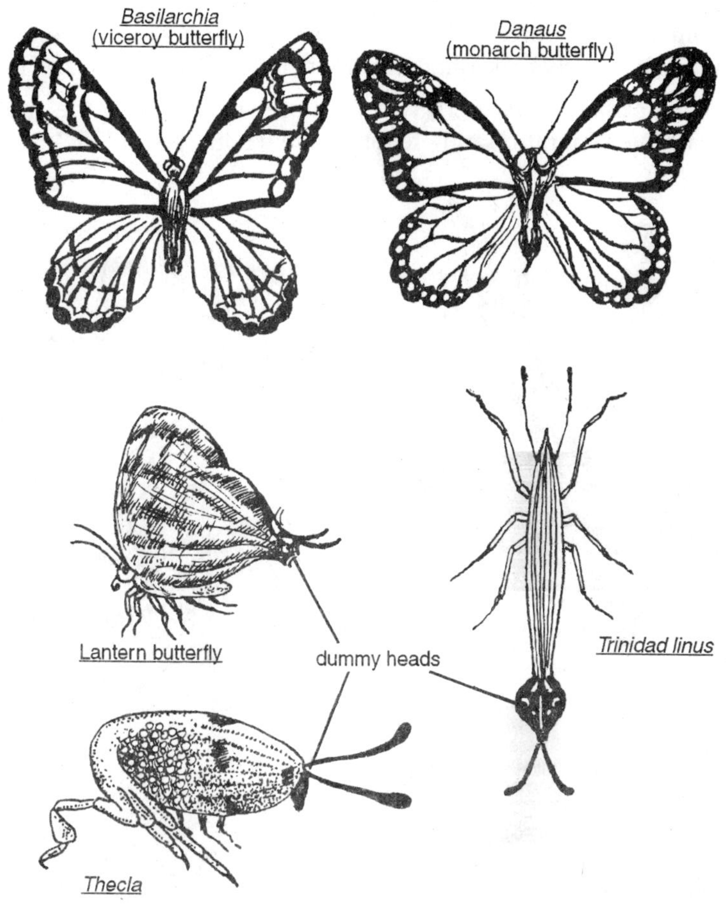

Fig.51 : Animals with Protective adaptations

butterfly *Basilarchia* and has the same reddish brown colour with black edges and also white dots. Thus, the viceroy butterfly can easily escape the birds. The hog–nose snake, *Heterodon*, behaves like a poisonous snakes by hissing and striking and also by flattening of the head. This snake confuses its enemies.

In *conscious mimicry* certain animals, on the apprehension of danger behave like dead bodies. So, they consciously become unconscious. For example, the American opossum, *Didelphis virginiana* rolls like a dead body and saves its life from the enemies.

Some hard–bodied beetles fall down like a pebble and lie inert when it is likely to be caught by its enemies. Thus, it saves its life.

In *Mullerian mimicry* two or more species have the same warning colouration and predators avoid both species. In *Batesian*

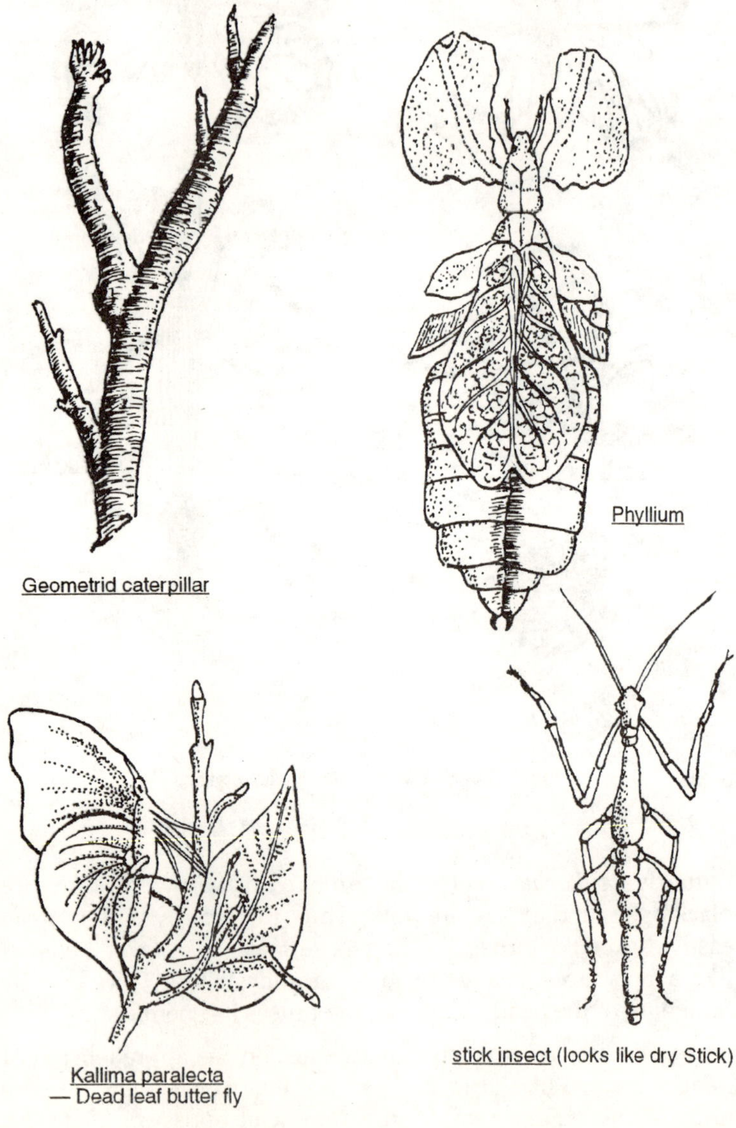

Fig.52 : Protective adaptations of some more insects

mimicry one of the two species is poisonous and distasteful. The other is eatable having good taste. But, both resemble each other and thus the mimic becomes disguised. Harmless and palatable insects mimic the poisonous and unpalatable species and both are avoided by the predators. Batesian mimicry was introduced by Bates (1862). Colours in general have a protective value for harmonising with the surroundings to evade the enemies. Certain colours aid in concealment and these render the animals inconspicuous. This type of colouration is termed *Cryptic.* For example white colour for an animal living on snow is a *cryptic* one. Signals are protective colours which comprise two types – *Recognition colours* and *Warning colours.*

Recognition colours are seen in gregarious animals. Members of herd can recognise their own kind and also opposite sex. Stripes of a zebra are both signal colour and protective marking. During the day time the stripes act as signals to others. At night they make the animal invisible in a background of shrubs.

Warning colours are also known as *Sematic colours.* They do no conceal the animals rather make them conspicuous. These colours can warn the enemies. These animals are inedible and dangerous. Blue spots on the body of *torpedo* advertise the awaiting dangers of electric shock. *Heloderma* the only poisonous lizard, moves with safety. Some animals can change colour by rapid dispersion of pigments in their *Chromatophores.* Chameleon has several layers of Chromatophores for colour change.

4. Animal Association Adaptations

There are different types and degrees of association between members of different groups or of the same group. These are —

(i) *Social or Communal living*: The bees, termites, waspes etc. live together in colonies. The bees are differentiated as Queens, drones and workers. Each one is adapted for particular utility. The queen lays eggs and is exclusively adapted for reproduction. The workers are adapted for performing large numbers of adaptations. The epidermis secretes wax, mouth parts are modified for collecting nectar and molding wax and the limbs are suited for collecting pollen.

Similarly, termites form large communities with a complex social system. There are workers, soldiers, complemental males and females and true males and females in the colony.

(ii) *Commensal adaptation*: It is an association of two animals in which one gets benefit and the other may or may not be benefited. For example, crabs live in pairs in the tube of *chaetopterus. Remora* (Sucker fish) attaches itself with the help of sucker to the body of shark mainly to get transport. An example of commensalism is seen between hermit crab and the sea anemone. The sea anemone uses the crab (or Gastropod shell) as a means of transport and the latter is protected from the predators by the stinging cells of the sea anemone. The participating members are called *commensals* and the association is known as *commensalism*. There is no physiological dependence here in this process. In each set of example given above one animal is benefited and the other is not harmed, rather, sometimes benefited. These animals can live independently also. The association is not compulsory.

(iii) *Symbiotic adaptations*: It is an adaptation between the two organisms of different species for mutual benefit. The life of the *symbionts* without each other is difficult. Presence of Bacteria in the intestine of rabbit is an example of symbiosis. Bacteria aid in the digestion and absorption of cellulose. Presence of *dinoflagellates* in the body of Hydra is also an example of symbiosis. Dinoflagellates in the process of Photosynthesis releases oxygen, which is utilized by Hydra, and CO_2 produced during respiration in Hydra is utilized for photosynthesis by dinoflagellates. So, there is physiological dependence between the two. One can not live independently. The association is known as *Symbiosis*.

(iv) *Parasitic adaptations*: Parasitism is an association between the two organisms one is known as the *parasite* and the other as *host*. The parasite lives at the expense of the host without paying anything in return. The parasite is benefited because it receives food and protection, but, the host is harmed. Several modifications are seen in the parasite. For example, structures for attachment (hooks, suckers, tentacles) develop and locomotory organs are lost. Other systems are reduced except the reproductive organs which are very much developed. The integument becomes thick and resistant in several cases. There are several parasites — *Entamoeba histolytica, Plasmodium vivex, Fasciola hepatica* and *Ascaris lumbriocoides* are good examples.

Concluding Remarks

Regarding the origin and evolution and mimicry, one of the views believes in chance similarities. Similarities between the organisms and the environment, there is possibility of chance resemblance among varied forms in nature. Weismann was of the opinion that the Natural Selection is the only factor responsible for the origin of mimicry. According to Fisher, the production of mimicry is because of dominant gene mutations. Whatever may be the reason for the origin and evolution of mimicry but it is a very important phenomenon which protects the organisms against their enemies. Self defence is an important factor for survival value. Thus, it is significant in evolution. Natural Selection, an important factor in evolution favours mimicry as it assures better survival and greater success in the multiplication of the race. If preadaptations are there, it becomes convenient for the organisms to explore some new adaptive zone.

Chapter 15

Fossils and Geological Time Scale

The word *fossil* literally means anything dug out of the earth. Remains of prehistoric organisms give the evidence of the presence of organisms which lived once upon a time. Fossils are defined as the *remnants of the living forms from the remote past.* These are geological records written in the rocks. They are preserved remains of entire animals and plants or their parts in the different layers of the earth or buried in the snow. Even the footprints of the animals well preserved are considered as fossils. The imprint of a leaf on the stone is also a type of fossil. The distribution of animals and plants in time, through a study of fossil records is studied in Palaeontology. It provides direct evidence of organic evolution. Thus, fossils are the curious objects abundant in coal beds, limestone beds or in common rocks. They are well preserved by natural causes.

Types of fossils and methods of fossilization

Remains of animals and plants are sometimes buried and well preserved by natural causes. Types of fossils and various processes of fossilization have been discussed in detail as under.

1. *Unaltered fossils*: When the conditions are exceptionally favourable the whole organisms or parts may be preserved in a continuously dry or cold regions of the world. Even soft parts of the animals are preserved intact and protected from decaying. Preservation of ancient elephants, *mammoths* in frozen ice in Siberia can be cited as an example of unaltered fossil. A ground sloth was found to be well preserved in the volcanic crater in New Mexico. Several insects have been found preserved in *amber* (fossil resin) exuded from the pine tree.

2. *Petrifactions*: Petrifaction is a very common method of fossilization. Dead bodies are buried in the sediments deposits on the floor of océans and other are water bodies. Some may be changed into stones if embedded in the rocks. *Sedimentary rocks* or *stratified rocks* are formed by the deposition of sediments under water. The original organisms undergo mineralization and molecule replacement between the organism and minerals like silica, calcium or iron takes place. The supersaturated solutions of minerals fill the cavities leaving the chemical composition of bones, shells and wood intact. As a result, the external forms and the histological details (*histometabasis*) are retained. The method of Petrifaction preserves both the hard as well as soft parts of the organisms.

3. *Moulds and Casts*: The mould is the impression or external morphology of an organism or an organic structure in the surrounding material (sediment). It is formed by hardening of the surrounding material or sediment or volcanic ashes in which the organisms was buried. Later on, the organisms decay and the organic material is removed by percolating water. As a result of this process, a cavity is formed and the external configurations of the organism become well marked but the internal structure is not obvious. The soft animals, like insects and jelly-fishes are preserved by this method. The cavity (or holes) in the body of the preserved animals is filled with mineral matters which later on harden to form perfect and *natural casts* of the original form. The moulds discovered are also filled with plaster of paris to produce *artificial casts*, thus giving the original shape to the fossil. Several natural moulds and casts are found in sandstones and limestones.

4. *Preservation of footprints or trails*: While moving on soft soil, wet sand and mud, the foot prints of moving animals are entombed and covered by sediments before they are disturbed. Thus the fossils of foot–prints are formed. Footprints of amphibians were discovered from Pennsylvanian period. *Trails* are also made below the surface of soft sediments and are filled with finer sediments. Trails also appear in the form of irregular markings due to crawling of certain worms or snails. It

may also be formed by the movement of fins of fish and dragging tentacles of jellyfish. The imprints as a whole provide shape, proportion and mode of progression of animals in the past.

In addition to the above mentioned types of preservations, several organisms are preserved in *Petroleum springs* and *sphalts, resins* and *ambers, ice* and *deserts* etc. The word Coprolite is used for the fossil preservation of the contents of the intestine and excrements of ancient animals. These are generally nodular and phosphatic.

Significance of Fossils

Fossils are of great importance in the study of evolutionary processes. They indicate definite lines of ancestry. Fossils are considered as the written documents of the process of evolution. They provide the following guiding principles :

1. *Evidence in favour of organic evolution*: The fossils of the organisms in the older rocks are simple in comparison to those of the new rocks of recent times. It indirectly points out towards the truth that complex forms of life have evolved from the simple forms. The succession of fossils helps in gaining knowledge of the ancestors of the present day organisms. One of the best known records of evolutionary changes are seen in the evolution of horse from the *Eohippus* to the *Equus* stage. A series of fossils have been obtained right from the *Eocene* time upto the *Pleistocene* and they have been critically analysed. Unbroken sequence of fossil records clearly show the Pedigree of certain animals groups (e.g. fossil of horse).

2. *Fossils and the Prehistroic life*: Fossils have authentic information about the organisms living in the past. For example, dinosaurs are known from their fossil records only. With the help of fossils, one can determine the relationship among various animal groups and evolutionary tree can be proposed.

3. *Fossils and Palaeogeography*: Fossils are helpful in reconstructing the ancient geography of earth. By the study of fossils, the nature of habitat can be depicted. For example, if the fossils of corals and echinoderms are obtained from a particular rock it can be presumed that the environment was marine there.

4. *Fossils as climatic indicator*: With the help of fossils, climatic conditions of that period can be determined. The distribution of seas and shores, mountains or the continental areas of the past can be indicated. Warmer climatic conditions and glacial or colder conditions are indicated by the presence of particular fossils.

Determination of Age of Fossils (Dating of Fossils)

There are some methods by which the correct age of the fossils by determining the age of the rocks (where the fossils are found) can be determined. The radioactive elements are found in the rocks. These elements change into non–radioactive isotopes at a fixed rate depending upon the half-life of the radioactive element. It may be pointed out that half life is the time required for 50 per cent disintegration of a given amount without any consideration of changing environment. The age of the fossils if determined accurately by *radioactive clock method* introduced by Boltwood (1907). Some radioactive substances are *Uranium, radium, carbon thorium, rubidium* etc. These elements disintegrate at a steady rate into a stable substance. Radioactive carbon (C^{14}) changes into ordinary carbon (C^{12}), Radioactive Uranium (U^{238}) changes into ordinary lead (Pb^{206}), Radioactive phosphorus (P^{40}) changes into non–radioactive argon (A^{40}), Radioactive rubidium (Rb^{87}) changes into non–radioactive strontium (Sr^{87}).

The breakdown of Uranium to lead and helium takes place as given below :

$$\text{Uranium}^{238} \rightarrow \text{Radium}^{226} \rightarrow \text{Polonium}^{218} \rightarrow \text{Astatine}^{218}$$

$$\downarrow$$

$$\text{Polonium}^{210}$$

$$\downarrow$$

$$\text{Lead}^{206}$$

$$\downarrow$$

1. *Lead method for calculation*: It is known that lg of Uranium yields 1/7, 600000000g of lead. Thus, if the original quantity of Uranium was m_1g and that of lead m_2g at the time of excavation, then the age of the rock can be determined in the following way :

   Quantity of Uranium (originally) = $m_1 + m_2$

   Quantity of lead (m_2g) formed by (m_1+m_2)g of

$$\text{Uranium in t years} = \frac{t\,(m_1+m_2)}{7600\,000\,000} = m_2 \text{ grams.}$$

$$\text{So, } t = \frac{m_2 \times 7600000000}{m_1 + m_2} \text{ years}$$

Thus, if the lead–uranium ratio present in the rock is known, it is easy to calculate the age of the rock and the fossils located there.

2. *Radio–carbon method*: This method is useful in determining the age of the fossils of late Pleistocene and recent periods. It is difficult to determine the age of very old fossils because the amount of radioactive carbon (C^{14}) decreases gradually with time. The living organisms contain a constant amount of heavy isotope of carbon i.e. C^{14}. In the dead and buried organisms C^{14} disintegrates into C^{12} at a definite speed. The half life of C^{14} is about 5568 years. By measuring the amount of radioactive isotope (C^{14}) still present in the fossil, the age of the fossil can be estimated by half–life method. This particular method has been suggested by W.F. Libby.
3. *Potassium–Argon method*: This method of calculating the age of the fossil is a recent one. Ordinary potassium contains nearly 0.01% radioactive isotope, which disintegrates with a half life of 1.3×10^9 years (i.e. 13000C0000) giving rise to calcium and Argon. By this method, the age of the earliest known fossil hominid remains of East Africa has been estimated to be 2000000 years.

Concluding Remarks

Fossil records are very significant in understanding the process of evolution. But, there is incompleteness of fossil records and there are some difficulties in fossilization. Most of the animals are destroyed by predators or Scavengers or are exposed to weather for disintegration. Soft parts of the body are not properly fossilized. Many fossils are crushed and distorted as they are located in the deeper rocks. Sliding of the earth's crust can disturb the sequence of fossils. Several fossils are unnoticed in the sedimentary rocks of the sea. Several places are not accessible for collection of fossils. Fossil records are thus fragmentary however, these records present a clear testimony of the fact of evolution.

The *Geological Time Scale* includes the history of earth ever since it was formed. The material forming the earth's crust which has been formed in layers over millions of years, is called *rock* by the geologists. There are three types of rocks in the earth's crust. They are *Igneous rocks, Sedimentary rocks* and *Metamorphic rocks. Igneous rock* is the oldest rock and was formed by solidification of molten material. *Sedimentary rocks* are formed from the materials moved by water and deposited as sediment under water. *Metamorphic rocks* are formed from the sedimentary rocks by constant heat and pressure. The rocks are arranged in layers (strata), one above the other, though this arrangement is often unrecognised by foldings and other irregularities. However, the strata can be arranged in series in which the oldest lies at the base and the most recent one at the top. Mountains and hills have been formed at many places as a result of folding and simultaneously the land masses moved to become higher or lower. The earth and all the living organisms have had a long history partly recorded in various rocks of the earth crust in the form of fossils. As estimated by geologists, the age of the earth is about 5000 million years ago or even more. The whole life span of earth is called the *geological time table*. The major divisions of geological time are called *eras*. These eras have been divided into *periods* which are further sub–divided into *epochs*. These *eras, periods* and *epochs* are arranged on the time scale in an order of their age. This arrangement is called "*geological time scale*".

The rocks having similar fossil forms have almost the same age. Rocks with simple fossils of plants and animals are older in comparison to fossils with more complex forms which are recent ones. The earth's crust made of strata was formed at different geological periods.

Fossils (remains of plants and animals) have been obtained from the different layers of the earth's crust. They are helpful in identifying rock strata and they indicate the chronological sequences. They show that various dominant groups of animals arose when great climatic changes took place. As we go higher in the strata of the earth, we find more complex fossils.

There are six major divisions of the geological time scale called *eras*. They are —

1. *Azoic era*: This was the earliest era. It was time of *igneous* rocks. There was absence of living organisms and hence there was no fossil.
2. *Archaeozoic era*: This era started about 1600 millions years ago. It was the time of initial life, living forms appeared first in this era. Sedimentation also started.
3. *Proterozoic era*: This era started about 900 millions years ago. This was the time of earliest life. Large amount of sediments were deposited.
4. *Palaeozoic era*: It started about 600 millions years ago. This was the time of ancient life.
5. *Mesozoic era*: It started about 270 millions years ago. It was the time of middle life. Land area attained almost the present form.
6. *Coenozoic era*: This era started about 60 millions yea 's ago. It is the present era and time of recent life. Coenozoic era is called the age of mammals.

More details of Geological time scale with notes on events in the evolution of life and environments have been stated in the Table.

Table

Geological Time Scale with Notes on Events in the Evolution of Animals with Changes in Environments

1	2	3	4	5	6
Era	*Period*	*Epoch*	*Time of Beginning of period in million of years (approx.)*	*Geologic and Climatic Conditions*	*Animal Life and Dominance of Particular group*
COENOZOIC (Age of Mammals)	Quaternary	Recent	0.025	Forests and deserts; warm climate; end of last glaciation	Age of modern man, development of human cultures; Rise of world civilization
		Pleistocene	1	Periodic glaciation resulted in destruction, unusual solar activity	Age of primitive man; extinction of many large mammals
		Pliocene	12	Cooler and temperate climate, Ice age, rise of mountains of America	Evolution of Man, Abundant mammals; Elephants; horses and camels like modern forms, Decline of some mammals
		Miocene	28	Some areas dry and arid; some areas cooler and wetter	Numerous mammals in their height of evolution; first man–like apes, modern carnivores
	Tertiary	Oligocene	39	Pleasant and warmer climate low land; continents began to have cooler weather later	Archaic mammals extinct, Modern mammals appeared

1	2	3	4	5	6
		Eocene	58	Zoned climatic belts were well established	Placental mammals diversified and specialized, hoofed mammals and carnivores well established. Archaic mammals disappear
		Palaeocene	75	Climate belts developed, climate warm but later on cooler	Extinction of great reptiles and evolutionary explosion of mammals
MESOZOIC (Age of Reptiles)	Cretaceous	—	135	Development of climatic diversity, warm climate but later on cooler	Dinosaurs reached their peak, became extinct; toothed birds became extinct, beginning of teleosts and modern birds; Development of snakes, Archaic mammals common
	Jurassic	—	165	Mild, Warm and humid climate throughout the world, plenty rainfall	Dominance of dinosaurs, lizards and crocodiles, appearance of first toothed birds, spread of giant reptiles, rise of insectivorous marsupials, some bony fish
	Triassic	—	205	Harsh and dry climate, mild to subtropical continents exposed	Extinction of primitive amphibians, transition of reptiles to mammals, rise of progressive reptiles and egg laying mammals, Rise of dinosaurs, bony fishes in sea.
	Permian		200	Continents uplifted, arid climate, glaciation in southern hemisphere, shallow sea	Extinction of Trilobites and ammonites, abundance of primitive reptiles, appearance of mammal–like reptiles, modern insects, decline of amphibians
	Pennsylvanian (carboniferous)	—	255	Uniformly mild climate, Glaciation and deserts also	Amphibians dominant, insects common, Rise of primitive reptiles

1	2	3	4	5	6
PALAEOZOIC (Age of Ancient Life)	Mississippian (carboniferous)	—	280	Desert conditions, climate uniform, huge swamp; humid and later on cooler, spread of tropical seas	Rise of insects, sea lilies at peak, spread of ancient sharks, rise of echinoderms in general
	Devonian	—	325	Local dryness uniform climates increased temperature, formation of coal, oil and gas	Ostracoderms end; fishes dominant group, diversification in fishes; shark and lungfishes abundant, evolution of amphibians, crabs, land snails, rise of cephalopods
	Silurian	—	360	Mild climate cooling, some areas arid	Ostracoderms, molluscs, expansion of invertebrates, first insects, rise of fishes
	Ordovician	—	500	Uniformly warm climate, volcanic activities at some places	First indication of armoured fish, corals abundant, diversified molluscs, decline of trilobites
	Cambrian	—	600	Melting of glaciers, warm climate, slow rise of sea levels	Invertebrates dominated; Rise of corals, Rise of shelled animals, Trilobites
PROTEROZOIC		—	900	Large amount of sediments, climate cool, volcanic erruptions; repeated glaciations	Direct fossil evidence lacking, shelled protozoans, coelenterates, flat worms, primitive annelides, soft marine invertebrates, fossil records rare
ARCHAEOZOIC		—	1600	Catastrophic and great volcanic activities, Little sedimentary deposition	Origin of unicellular life; No. fossil forms, indirect evidence of living things from deposits of organic materials in rocks

Chapter 16
Animal Distribution

Life occurs in all habitats ranging from very high (even more than twenty thousand feet) mountains to the deep sea areas upto the depth of nearly 10000 metres. But, all the animals are not present in all habitats, rather, they are confined to certain particular areas over the globe which are known as *range*. Thus, range is supposed to be the area of distribution of a species.

There are two aspects of distribution of animals —

1. Distribution of animals in space.
2. Distribution of animals in time.

Distribution of animals in space : That the animals are distributed over the surface of the earth both on land and in water. It deals with the horizontal distribution of animals in different continents and islands. The study of *geographical distribution* of animals is called *zoogeography* and these realms are known as *Zoogeographical realms*. Another type of distribution in space is the vertical distribution of animals on land and water both. It includes *holobiotic* (marine), *limnobiotic* (freshwater) and *geobiotic* (terrestrial) *realms*. It is known as *Bathymetric distribution*, which has been discussed later on in detail.

Distribution of animals in time (geological distribution) : It covers the evolutionary history of the animals, dealing with the distribution of animals in the past, based on fossil records.

Patterns of Animal Distribution

There are three different patterns of animal distribution. These are —

1. Continuous distribution

2. Discontinuous distribution,
3. Bipolar distribution.

Some animals are cosmopolitan having wide range of distribution. They are found all over the world. For example, mytilus, bats, rats, cuckoos etc. are present almost everywhere in the world. Such distribution is known as *Continuous distribution*. When the distribution is uniform it is referred to as *eurytopic* but when certain animals are confined to particular areas only, it is known as *stenotopic*. Adaptations of the eurytopic forms are generalised where as the stenotopic forms have specialised adaptations.

When there is no continuity in the distribution of the individuals of the same species or related species over the globe, the distribution is known as discontinuous. They are found to inhabit some localised areas. For example, *Protopterus* is found in Africa, *Lepidosiren* in South America, *Neoceratodus* in Queensland (Australia), *Ostriches* in Africa and Arabia, Emus in Australia and so on.

In the *Bipolar distribution*, the distribution is localised to polar regions (North and South Poles) only. Several species are found inhabiting only the arctic and antarctic waters and no where in the other parts of the world. For example, *Sardina* and *Botrynema* among the coelenterates and *Oncorhynchus* among the fishes are found on poles.

There are several factors which hinder in the normal distribution of animals. These factors are known as barriers. There are three categories of barriers —

1. Physical barriers
2. Climatic barriers
3. Biological barriers

The *Physical barriers* include topographical barriers like Himalayan range in India and Alps in Europe. They become obstacles in the continuous distribution of animals. Large bodies of waters and land masses, salinity of water, vegetative barriers and large distances are also Physical barriers. *Climatic barriers* are temperature, moisture and amount of light which control the dispersal of animals. The *biological barriers* include food and enemies. Sedentary habit and territorial weakness of some animals also limit their distribution.

Bathymetric Distribution (in detail)

It has already been pointed out that *bathymetric distribution* is the vertical distribution of the animals in space both on the land and in water. This sort of distribution takes place from the highest Alpine peak, to the abyssal depths of the sea. Abyssal zone receives no light. There are great contrasting conditions between the two which affect animal distribution. The vertical distribution of the animals is affected by certain Primary and Secondary conditions which have been described below :

Primary Conditions

It has been discussed below under three different heads.

1. *Air or Water medium*: That the purely water breathing forms can not survive in air. Similarly, purely air breathing forms can not live in water. Of course, there are some animals which are both water breathers and air breathers (e.g. *Rana tigrina*). They can live for considerable period outside their original habitat. Temperature of water controls *phytoplankton* and *zooplankton* densities.

2. *Presence or absence of light*: It there is no light and there is complete darkness, the living fauna will be completely modified for cold and dark and quiet environment. Presence or absence of light affects the animal population through the food supply. There are variations in the life processes due to rhythms of light and darkness. It is a fact that Rhythmicity is genetically determined, but the organism adapts or adjusts its activity–rest rhythm to the altered times of light and darkness. The process of Photosynthesis in deeper water of the sea occurs with the blue and green rays which are absorbed by the brown and red pigments of red algae (*Phyllophora*). Fauna of a particular zone depends on the nature of the food available.

3. *Presence or absence of substratum*: Nature of substratum is one of the ecological factors which governs the distribution of animals and plants in the sea. The fauna and flora are definitely influenced by the nature of the bottom of the zone which may comprise hard rocks, pebbles, gravel, sand or soft mud etc.

   If there is absence of substratum, the animals should be *buoyant* or *able to swim*. Only then they can change the place. The bionomic group of the animals is determined by the above factors.

The ocean has been divided into several ecological zones according to different depths. These are —

(i) *Intertidal* or *Littoral zone* – Which is present on every shore and is exposed to the tides. It is most diversified and is richest in benthos.

(ii) *Continental shelf* : Which is present beyond the littoral zone mainly consisting continental islands. Light penetrates for a depth of 400 feet, adequate for synthesis.

(iii) *Neritic zone* : Which are shallow water near the shore. This zone is found over continental shelves and it is not over 600 feet deep.

(iv) *Pelagic zone* : It is the open water of the sea. The upper part of waters is known as *epipelagic* (200 metres).

(v) *Bathyal zone* : It is the zone in water upto 6000 feet deep and is scantily inhabited.

(vi) *Abyssal zone* : It lies below the bathyal zone and there is no light.

Terms used for some aquatic forms

(a) *Benthonic organisms (Benthos)* : Those living in or on the bottom. They are mostly sessile forms e.g. chiton, oysters, sponges, crabs etc.

(b) *Plankton* : These are minute floating organisms usually found in surface waters. Its movement depends upon the direction of water currents. They have been divided into categories —

Nannoplankton : They are less than 0.06 mm in diameter (e.g. Bacteria and Protozoans).

Microplankton : They are less than 3 mm and more than 0.06 mm (e.g. distoms and dinoflagellates).

Macroplankton : They are larger than 3 mm in diameter (e.g. Jelly fishes).

(c) *Neuston* : These are floating forms at the surface of the shallow waters.

(d) *Bathy–pelagic* : These are organisms inhabiting deeper water.

(e) *Nekton* : These are larger swimming or flying animals. They have great power of locomotion. Large crustaceans, molluscs, fish, turtles, sea birds, whales, seals etc. are aquatic forms. Insects, bats and the birds are the aerial nektons.

Secondary Conditions

There are some secondary conditions also which limit the vertical distribution of animals. *Fresh* or *Salty* water can act as a secondary condition to limit the vertical distribution of animals. With the increase of pressure with the depth the water dwelling forms are affected much. The animals exhibit unique adaptations. Animals may be dark or red in colour with sensitive eyes or receptors. They may exhibit bioluminescence in the dark surrounding.

Organic Realms

On the basis of bathymetric distribution of animals there are three organic realms. These are

1. Geobiotic or terrestrial
2. Limnobiotic or fresh water,
3. Halobiotic or marine

Geobiotic realm extends from high tide–mark along the shores of all the continents and islands up to the highest possible elevation (Alpine peak). In between these high tide mark and alpine peak several contrasting differences can be observed in the climatic conditions. Altitude and climate are supposed to be important factors in the distribution of both plants and animals. Cryptozoic or subterranean caves may be considered as Geobiotic sub–realm. There is absence of light in the caves and as such marked changes can be seen in the body organization of the permanent cave dwellers. Cave adaptations have been discussed in "Adaptations" chapter of this book.

Water, found on the surface of the land i.e. in rivers, streams, lakes, ponds and reservoirs is known as inland water. *Lentic water* is the standing water and *lotic water* is the running water. Habitats of these water are respectively called the *lentic habitats* and the *lotic habitats*. There is limitation of fauna in these habitats which come under *limnobiotic realms.* There is seasonal variations in the animal and plant population in the freshwater realms.

Halobiotic or marine realm is a very important realm. Several contrasting characters are present in the animal of this realm. There has been sufficient adaptational changes among the various groups of animals in the sea. The marine realm cane be conveniently divided into four sub–realms — e.g. *Strand* or *tidal zone, Flat sea* or *Shallow sea zone, Pelagic zone* and *Abyssal zone.*

Strand or *tidal zone* is characterised by (i) penetration of light and (ii) presence of substratum. It covers the transitional area between the terrestrial and marine realms. The animals come to the shore alongwith the tides but the receding tides leave them on the sand. They protect them from drying with the help of shells or other devices. Some of them burrow in the moist sand. They can breathe air or water, or both. The area of the tidal zone varies greatly.

In *Flat sea*– or *Shallow sea zone* also, light and substratum are necessarily present. It is *continental shelf* covered by the water below low watermark. The sloping continental shelf extends from the coast line to about 16 km. The angle of descent undergoes abrupt change and becomes a steep slope called the *continental slope.* The *benthonic* organisms can live conveniently in this sub–realm (shallow sea zone). This sub–realm is known as the "cradle of evolution" as all the factors necessary for evolutionary process are present there (e.g. light, suitable temperature and proper vegetation). The *flat sea* is very rich in its fauna. Animals like foraminifers, sponges, sea anemones, corals, nereis, aphrodite various enchinoderms, bryozoans, crabs, mussels, amphioxus, cuttle fish, sea snakes, otter etc. are in abundance.

Pelagic zone consists of water up to the depth of about 600 feet. Light can penetrate in this zone but there is absence of substratum. The upper part of this zone the temperature is higher and the wave is frequent and violent. But, in the lower part of this zone, the wave is very slow and the water temperature is greatly reduced. *Nectonic* or *Planktonic* forms are in abundance in this zone. Some fishes, whales and porpoises, petrels and floating plants are also found.

Ecological zones in different oceans will have different fauna and flora when they are separated by barriers. East and West coasts of America have been separated by vast land and as such, there is difference in fauna and flora. Below the pelagic realm, lies the *Abyssal realm*. It receives no sun light. This realm has been divided into *Abyssopelagic zone* and *Abysso–benthonic zone*. In the former there is no substratum and the animals can only swim or float but in the latter zone the substratum is present and some animals are attached to it. Light may be present in transitional abyssal layer only but in the deep layer there is complete absence of light. The Animals are mostly carnivorous. As we go deeper into the *abyssal zone*, we find increasing pressure of water. The bottom layer is extremely cold and perfectly quiet. From evolutionary view points this zone is not fit. Bioluminescence is exhibited by some animals in the dark layers.

Chapter 17
Evolution of Horse

Among the vertebrates, the phylogeny of horses is most completely known. The phylogeny of horse is considered as the best palaeontologcial evidence in support of organic evolution. The evolution of horse initiated in the early Eocene period. It is a case of progressive evolution where divergence took place from the main line from time to time. Modern horses (*Equus equus*) evolved from *Eohippus*, a small creature of only 12 inches height.

Factors Initiating Horse Evolution

The important factors which may be considered responsible for initiating horse evolution are the following :

1. The rainfall was greater,
2. The forest was dense and luxuriant,
3. They had to feed on succulent herbage,
4. They found shelter behind the bushes against their enemies.

But, later on the climatic conditions changed and the inferior planes began to rise. The forests were diminished and grassy planes started developing.

Equine Adaptations

The adaptations undergone by the horses are mainly for two purposes —

(i) Perfection of food getting mechanism

(ii) Perfection of the mechanism of speed.

In response to the first adaptation, the conversion of pre–molars into molars, appearance of lophodont and solenodont and development of cement substance for strengthening the teeth took place. Similarly in response to the second adaptation, change from plantigrade to unguligrade, lengthening of the head, neck and limbs, reduction of digits, perfection of hoof, absence of canines or tusks and formation of a long diastema were seen. Large brain and high intelligence are also considered as equine adaptations.

Stock and Place of Origin

Horses originated from *Condylarthra*, a group of primitive five toed ungulate. Two allied genera i.e. *Hyracotherium* in London (Europe) and *Eohippus* in America were derived from this stock of ungulate, now extinct. There are evidences that *Hyracotherium* became extinct earlier and as such American genus *Eohippus* has contributed much.

Stages of Horse Evolution

Various stages in the evolution of horse have been described below with important features —

1. *Eohippus*: Its fossils were procured from Eocene beds of North America particularly from Big Horn Basin in Northern Wyoming and the San Juan Basin in North–Western New Mexico. The important characters were —

(i) They were forest dwellers,

(ii) They were about 12 inches in height,

(iii) Skull and neck were short,

(iv) Legs were digitigrade, the ulna was flexible and separate from radius, fibula was stout and separate from tibia.

(v) Four toes (2nd, 3rd, 4th and 5th) were present in the forelimb, the 1st toe being absent.

(vi) In the hind limb, only three functional toes (2nd, 3rd and 4th) were present, first and fifth being absent.

(vii) 44 teeth were present there being a tendency of diastema formation. Molars and pre–molars were different, teeth were low crowned (*Brachydont*)

(viii) All the toes touched the ground and were supported by pads.

(ix) They were browsers.

2. *Orohippus*: Fossils were procured from North America and New Mexico during the Eocene time. They were known as mountain horse. The important features were —

   (i) There was slight increase in the size and they attained a height of 13½ inches.

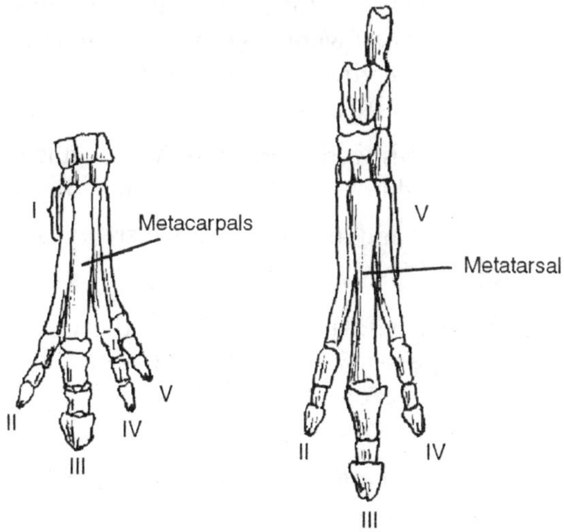

Fig.53 : *Eohippus* (hand and foot bones)

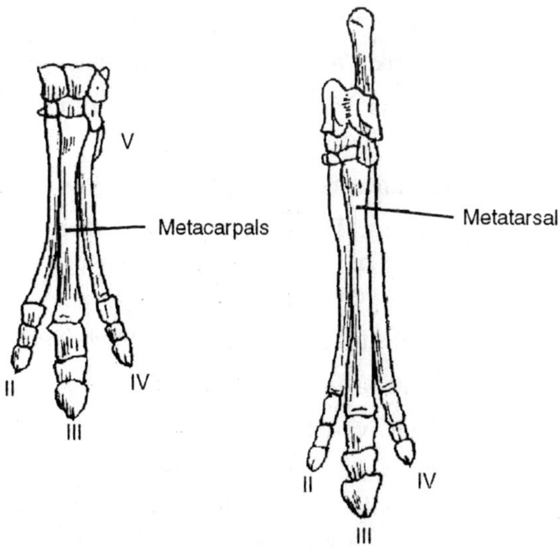

Fig.54 : *Mesohippus* (hand and foot bones)

(ii) There were 4 toes in the fore limb and 3 toes in the hind limb, but tiny vestiges of first and fifth toes were lost.

(iii) Middle digits in both limbs were dominant.

(iv) The last pre–molar (IV) was just like molar in form.

(v) They were still browser.

3. *Epihippus*: The fossils were procured during the Eocene time. They were slightly larger than *Orohippus*. They became extinct by the end of Eocene. The important characters are —

(i) 4 toes were present in the forelimb and 3 toes in the hind limb. Vestiges of I and V digits were absent.

(ii) 3rd and 4th pre–molars were molariform.

(iii) They were browsers.

4. *Mesohippus*: They were directly derived from the *Epihipus*. The fossils were discovered from the Big Badlands of South Dakota during the Oligocene period. Important characters are —

(i) The size increased upto 24 inches and they were about the size of the sheep.

(ii) Feet had 3 functional toes with pads under them.

(iii) Equus like head, short and flexible neck.

(iv) Trunk was long and slender, back was more arched.

(v) It was adapted for rapid running (Part below knee and elbow elongated).

(vi) Ligaments developed for rapid running.

5. *Miohippus*: They were present during the Oligocene period. They were slightly larger than *Mesohippus*. The important characters are —

(i) Three toes were present, which were broad.

(ii) Teeth were low crowned.

(iii) They were browsers

(iv) The cannon bone (Metatarsal) in both the limbs were in contact with outer ankle bone.

6. *Anchitherium*: They flourished during the *Miocene* period. Fossils were procured from North America, Europe and Asia. Important characters are —

   (i) They had three toes,

   (ii) They were close descendent of *Miohippus.*

   (iii) Cheek teeth were simple.

   (iv) They were forest dwellers and were slightly larger than *miohippus.*

7. *Hypohippus*: They evolved from *Anchitherium* during *Miocene.* The fossils were obtained from North America. They later on migrated to Eurasia. In Pliocene they were recorded from China also. The important characters are—

   (i) They were larger in size (about 40 inches high) in comparison to *Anchitherium.*

   (ii) They had three toes.

   (iii) Teeth were broad and low–crowned.

   (iv) They were browsers on succulent herbage.

   Megahippus of America was an offshoot of *Hypohippus.* It had spoon like lower incisors pointing forward and it was slightly larger than Hypohippus in size.

8. *Archaeohippus*: They are known as Pigmy horses of *Miocene.* They were smaller in size than *Miohippus.* They might have been derived either from *Miohippus* or *Parahippus.*

9. *Parahippus*: They were present during the Miocene period and were on the direct line of equine evolution. They descended from *Miohippus.* The important characters are—

   (i) They were three toed

   (ii) The teeth cheek became adapted for grinding. Teeth were low crowned (hypsodont).

   (iii) Cement substance was laid down on the enamel of the teeth in the progressive forms.

   (iv) The third digit in the forelimb and hind limb became more predominant.

   (v) The median digit was the most effective in walking.

   (vi) They were browsers.

10. *Merychippus*: They were present between the middle and upper *Miocene* period. They directly evolved from *Parahippus*. The important characters are —

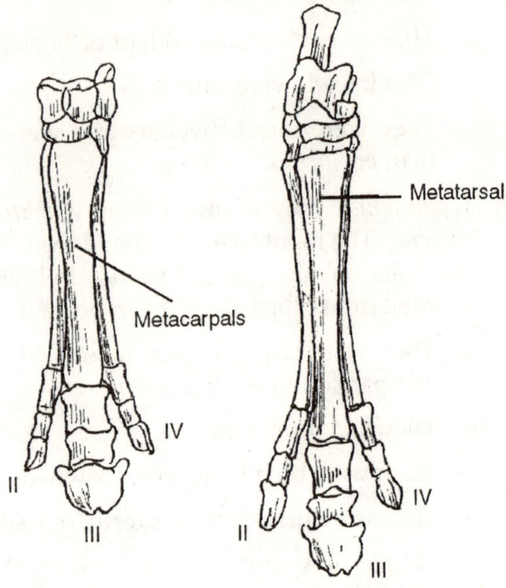

Fig.55 : *Merychippus* (bones of hand and foot)

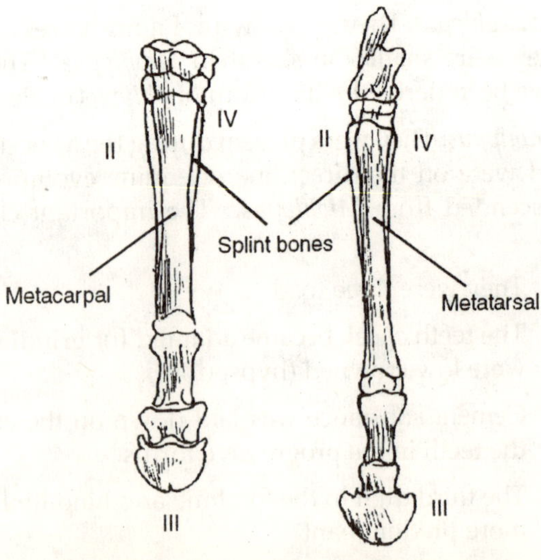

Fig.56 : *Pliohippus* (bones of hand and foot)

(i) They were about 40 inches high

(ii) Milk teeth were short crowned having very little or no cement at all.

(iii) Permanent teeth were high crowned and heavily cemented suited for grazing habit. It was a transitional period from browsing to grazing habit.

(iv) The skull became almost *equus–like*.

(v) Ulna fused with the radius.

(vi) The fibula was represented by a functionless spike of bone projecting in downward direction from the head of the tibia.

(vii) The feet retained three toes, the side toes (II and IV) were reduced and they were not touching the ground.

(viii) Ligaments of the muscles were greatly developed.

(ix) The central toe ended in a convex hoof as in *equus*.

(x) Pads under the toe were absent.

11. *Protohippus*: They evolved from *Merychippus* during the later Miocene period. The important characters are —

(i) Teeth were long crowned and heavily cemented.

(ii) They acquired grazing habit.

Protohippus gave rise to *Hippidion* which was present in *Pliocene*, from which *Onohippidion* evolved during the *Pleistocene* time. In the *Pleistocene* period *Hypohippidion* also evolved and became extinct.

12. *Hipparion*: Fossil remains of Hipparion were discovered from different parts of Asia and Europe and from also a part of Africa. They were present during the early *Pliocene*. They evolved from *Merychippus*. The important characters are —

(i) They were three–toed grazers.

(ii) They possessed high crowned grinding teeth.

(iii) It was about 40 inches in height.

(iv) They were great travellers.

Hipparion gave rise to *Stylohipparion* in Africa during the *Pliocene* which became extinct.

13. *Neohipparion and Nannippus*: They were present during the middle Pliocene period. They evolved from *Merychippus*. The important characters are —

   (i) *Neohipparion* was of low bony (small horse) size,

   (ii) *Nannippus* was a pigmy creature,

   (iii) *Neohipparion* became extinct in the upper Pliocene but *Nannippus* survived and became extinct during the lower Pleistocene.

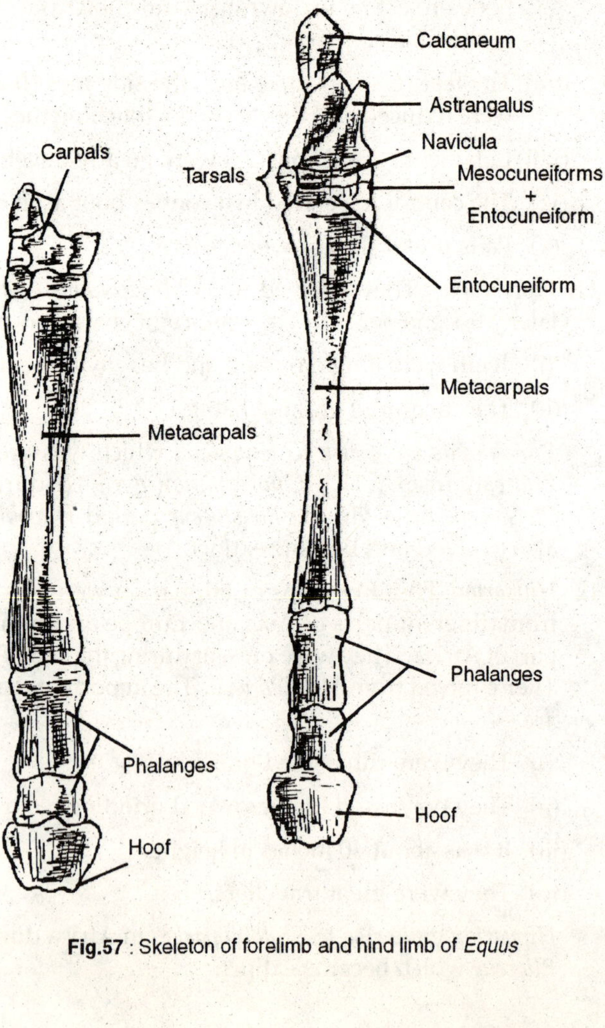

Fig.57 : Skeleton of forelimb and hind limb of *Equus*

(iv) Like *Merychippus* they retained the side toes,

(v) Teeth were curved inward in both the cases.

14. *Calippus*: It evolved from *Merychippus* during the middle of *Miocene* and became extinct during the lower *Pliocene*. The important characters are —

 (i) It was quite small in size (Pigmy),

 (ii) It had slightly longer teeth in comparison to *Merychippus*.

15. *Pliohippus*: They were present in the upper *Pliocene* period. They were popularly known as the *Pliocene horse*. The important characters are —

(i) It was 40 inches in height.

(ii) It was the first one–toed horse.

(iii) Teeth was longer than *Merychippus*.

(iv) A peculiar pit was present in front of the orbit.

(v) The side toes (2nd and 4th) were reduced.

(vi) There was increase in the size and complexity of molars.

Pliohippus was succeeded by the modern horses (Genus – *equus*).

16. *Plesippus*: Fossils of *Plesippus* have been obtained from Texas during the middle of *Pliocene* period. They gave rise to more advanced horses. Important characters are—

 (i) They are like modern horses of the size of Arab horse.

 (ii) Teeth had less curved crowns.

17. *Equus*: *Pliohippus* gave rise to *Equus* representing the modern horses and their allies. The equine adaptations have already been discussed in this chapter. *Equus* appeared in the late *Pliocene* and they achieved worldwide distribution during *Pleistocene*. It is about 60 inches in height and it has lost the first and fifth digits but the second and fourth digits are represented (splints). The weight of the body is balanced by the third toe only. The hoof is well developed. Complicated enamel ridges are present on the elongated crowns of the molars. Legs are quite long and the brain is well developed. It is well adapted for swift running and feeding on all sorts of grasses and grains.

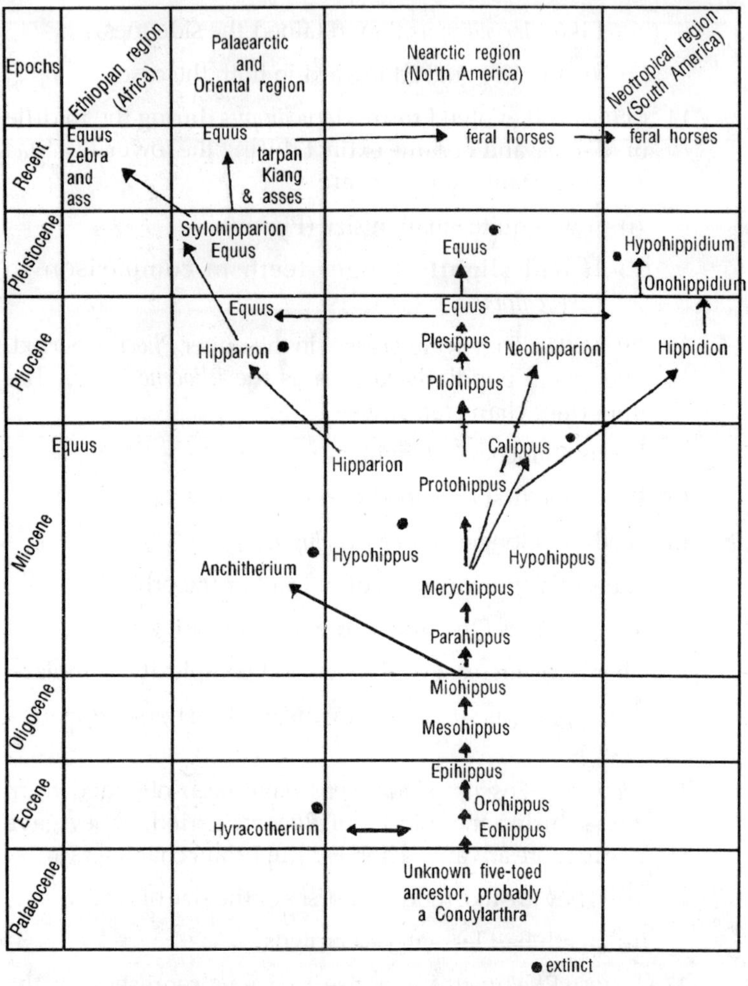

Fig.58 : Evolutionary tree (Phylogeny) of horse

Conclusion

No further horses appeared in Europe until the *Miocene* period as the genus *Hyracotherium* became extinct already. Only the N. American genus *Eohippus* contributed much. Today the important modern horses include —

(a) *Equus equus* – True horses

(b) *Equus hemionus* – Asiatic Wild ass

(c) *Equus caballus* – Wild horse

(d) *Equus caballus perzewalskii* – Mangolian wild horse

(e) *Equus asinus* – African Wild ass

(f) *Equus zebra* – Mountain Zebra

(g) *Equus burchelli* – Common Zebra

Chapter 18

Evolution of Elephant

Elephants (*Elephas*) belong to order *Proboscidea* of Class – Mammalia. The evolution of elephant has been traced back to the Eocene and Oligocene epochs. They have evolved from the *Proboscideo–sirenian* ancestors. Fossils have been obtained from Asia, Europe, Africa and America. During the Oligocene period, four principal stocks from Proboscidea have been identified. They were *Moeritherium* (having 36 teeth), *Dinotherium, Mastodons* and Elephants. Mastodons included – (i) *Trilophodon* (with three crests in intermediate molars), (ii) *Tetralophodons* (with four crests in intermediate molars), (iii) *Dibelodons* (without lower tusks), and (iv) *True Mastodons* (with shorter lower jaw).

Finally, the elephants had two important forms —

(i) The *Stegodonts* – transitional browsing elephants.

(ii) *Elephas* (true elephants) – having only lower tusks, reduced lower jaw and complex grazing teeth.

Elephants probably originated from the fossil genus *Phiomia* during the oligocene period. This genus had 28 teeth only. Phiomia is supposed to be the second succeeding genus from *Moeritherium* in the evolutionary series.

In *mastodons* there were 5-6 crests and in *Stegodon* the number increased upto 10. In the wooly mammoth (*Mammonteus primigeneus*) the number increased upto 25-30. There are only 10-12 crests in *Loxodonta*. The deep crowned complex grinding teeth are present in *Elephas*. It has been observed that there is a gradual increase in the size of the upper tusks from *Moeritherium* to *Stegodon* and *Elephas* but the lower tusks are lost, and the lower jaw shortened. The proboscis gave rise to long trunk during the course of evolution. Now the Polyphyletic theory of Osborn seems to be more convincing. According to this theory, three different lines of Proboscidian evolution took place from the ancestral *Moeritherium*. They were *Dinotheres, Mastodons* and *Elephantoids*. The modern genera evolved from the last offshoot.

More about Fossil and Living Elephants

Asia is supposed to be the true home of elephants. The evolution of true elephants has been traced back to the form *Stegomastodon latidens,* which was discovered from the upper Miocene period of Southern India. It gave rise to *Mastodon elephantoides* or *Stegodon clifti.*

Fossils of several species of wooly mammoth e.g. *Mammonteus primigeneus* were found in Europe and North America during the Pliocene and Pleistocene times. The American species included *Elephas imperator* and *E.jeffersoni, E.antiquus* was discovered from Europe. They were primitive and possessed straight tusks. They exhibited some affinities with *Loxodonta*.

At present, there are two very important species of living elephants. They are *Elephas maximus* and *Loxodonta africana,* the former living in India, Sri Lanka, Burma, Thailand, Sumatra and Borneo and the latter living in Africa mainly, South Sahara and some other places. The Asiatic elephant *Elephas maximus* has five hind toes but the mammoths have only four toes. Also, in mammoths, the tusks are spiral. *Loxodonta africana* has enormous ears, lower forehead and large tusks. They also possess greater height in comparison to other living elephants.

Modern elephants are characterised by —

1. Large and massive body with pillar like limbs.
2. They may be 11 to 13 feet in height and 6 to 7 tonnes in weight.
3. The skull is mechanically adapted to adjust the leverage for the weight of the trunk.
4. The limb–bones are perpendicular, the ulna dominates in the fore limbs and radius is reduced.
5. Five toed Plantigrade limbs are present.
6. The skull is mechanically adapted to adjust the leverage for the weight of the trunk. The trunk is known as

proboscis. The nostrils terminate at the tip of the proboscis and compensates for the short neck.

7. There are only 28 teeth and dentition is that of *lophodont* type (molars having crests).

8. The second pair of incisors are modified as a pair of tusks.

The fossil Proboscidians appeared not only during the Eocene and Oligocene periods but also during the Miocene and Pliocene times. In the Eocene and Oligocene periods, the important fossil forms were – *Moeriherium, Phiomia* and *Palaeomastodon*. They have been described below :

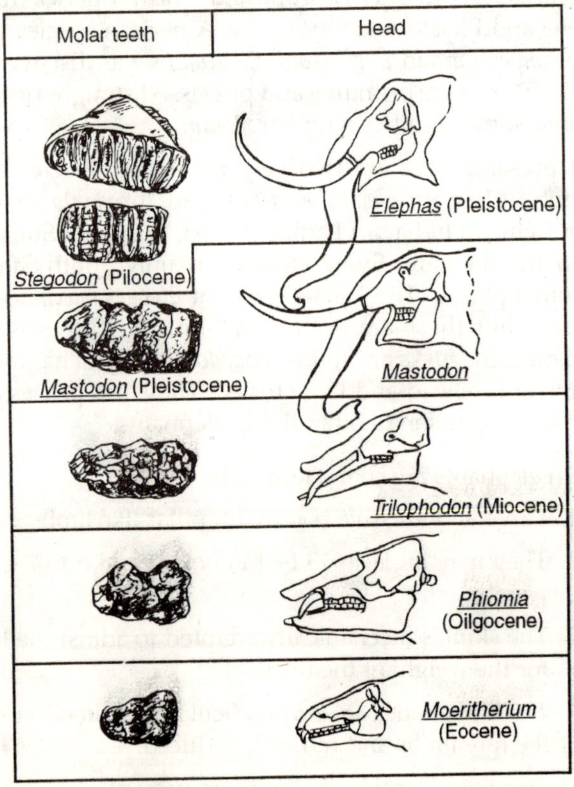

Fig.59 : Evolution of head and molar teeth of Mastodons and Elephants

1. ***Moeritherium*** **:** It was discovered near the lake Moeris in Egypt. It had the following characters —

i. They were swamp dwellers, about 3 feet in height.

ii. They were about the size of a Pig.

iii. There was no indication of Proboscis.

iv. The tusk formation was initiated from the upper incisors.

v. The nasal openings were large.

vi. Nostrils were present behind the tips of the snout.

vii. Molars were small and short crowned with two simple transverse crests.

viii. The lower jaw was elongated.

ix. There were 36 teeth in all.

2. ***Phiomia*:** The fossils were discovered from Moeris in Egypt. Recently, the fossils have also been found in Sewalik hills (India). The important characters are —

i. Neck was long and the lower jaw was elongated.

ii. 2nd pair of incisors (upper jaw) were modified into downwardly curved tusks.

iii. Upper and lower canines and incisors (except 2nd pair) disappeared.

iv. They were larger in size almost double than *Moeritherium*.

v. Molars had three transverse ridges (*trilophodont*).

vi. Limbs were basically like modern elephants.

3. ***Palaeomastodon*:** They were primitive mastodons discovered during the middle Oligocene and they gave rise to true Mastodon (*Mastodon americanus*). They were forest living creatures with short jaws. They were about 6 feet in height. The teeth were simple.

During the Miocene and Pliocene the following fossil forms were formed. They have been described with important characters (below) :

1. *Dinotherium*: The fossils were obtained from Europe and India. They lived during the Miocene time and became

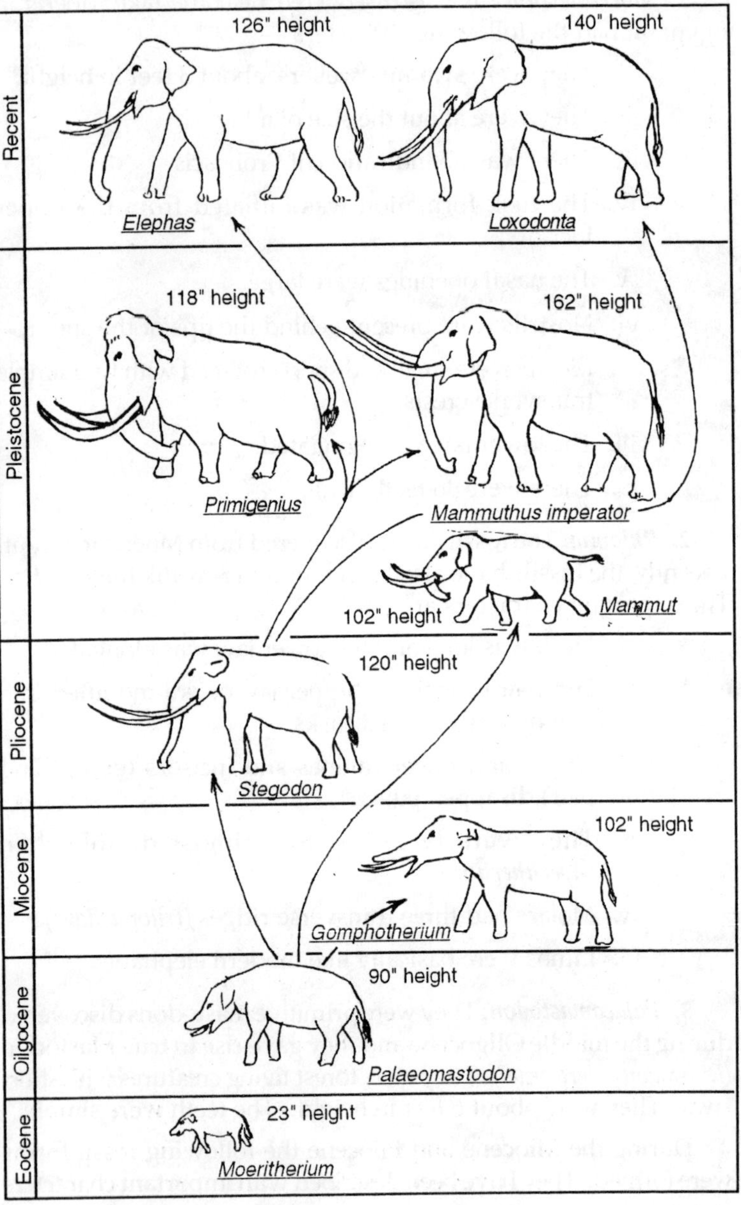

Fig.60 : Evolution of *Elephas* and *Loxodonta* from *Moeritherium* (Changes in height and trunk to be marked)

extinct in pliocene. The diverged from the main Proboscidian stem. The important characters are —

(i) They had large body and long limbs

(ii) The upper tusks were lacking

(iii) The lower jaw had long tusks, and abruptly bent downwards at the symphysis.

(iv) The trunk is well developed.

(v) Lower tusks were used for digging the ground and extracting roots in plants.

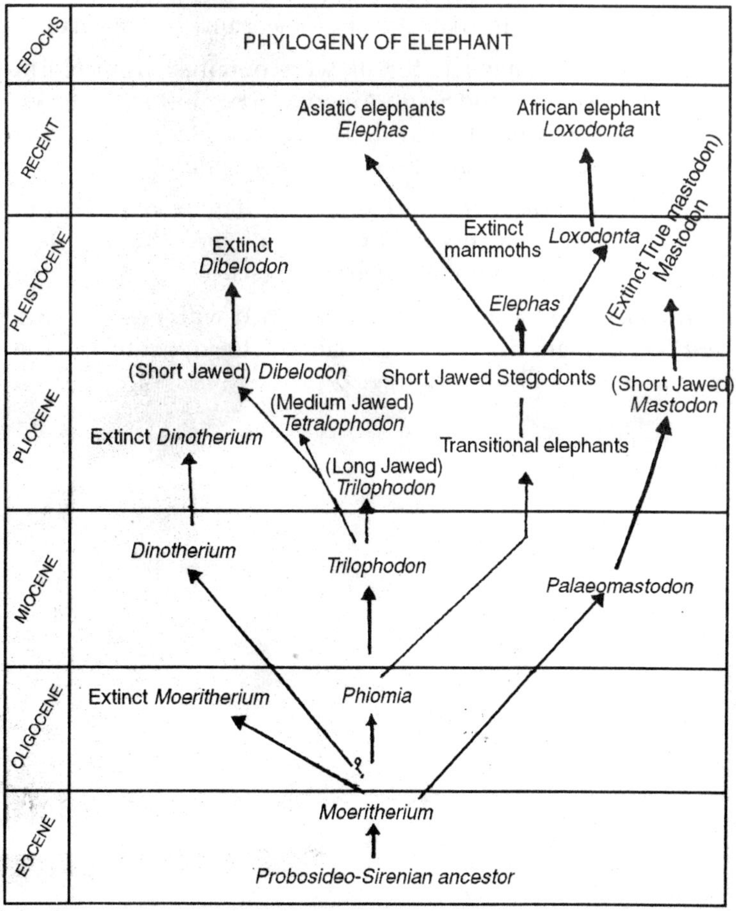

Fig.61 : Evolutionary tree of elephants

2. *Tetrabelodon*: They included four–tusked mastodons with elongated lower jaws.

   (a) *Trilophodon (Gymphotherium)* : They probably migrated from Asia to North America. They had three crests in the intermediate molars. The upper lip was elongated into long mobile proboscis. The upper tusks were slightly curved. Six cheek teeth were present in the upper and five in the lower jaw. Example is *T. angustidens*.

   (b) *Tetralophodon*: They were discovered from Italy during Pliocene. Jaws were not much elongated but the grinding teeth were present. 4 crests were present in the intermediate molars – lower tusks were small.

   (c) *Dibelodon*: The fossils were obtained from North America and South America. They lost lower tusks. Jaws were shortened. They were replaced by the true mastodons in the southern hemispheres.

   (d) *Mastodon*: During Pleistocene, they were present in Europe across Asia to Alaska, USA. Molars were simpler and there were 5-6 crests.

In males, the lower jaws were shortened with tusks. Males were 7-9 feet high. There was no enamel in the upper tusks. The contour of the skull was primitive.

Chapter 19
Evolution of Camel

Camels popularly known as the "Ship of the desert" belong to group *Artiodactyla* of the order *Ungulata* (Class – Mammalia). They are represented by 34 species in America and nearly 250 species in Africa and Eurasia. The *camelus* and the *Llama (Auchenia)* are the two living genera, the former is confined to the old world but the latter is present in the new world. In the case of *Llama* (South American genus), there is absence of hump and they are short statured and covered with wooly hairs. Genus *Camelus* is however characterised by the presence of hump. Several fossil remains of camels are available which reveal its evolutionary history.

Characters of Living Camels

There are several important characters of the living camel which have been basically acquired for living in the deserts. They have been enumerated below :

1. The head is projected upward just to protect the eyes and nostrils from heat reflected from sand of the desert. Also, sand particles do not enter easily into the eyes and nostrils. Eyes are protected by hairs and eye lashes and the nostrils have flaps.
2. The hump on the back is a storage for fat which is a source of nourishment when there is scarcity of food in the deserts.
3. The speed of the camel is very fast in the desert. The feet are provided with one or two cushion like pads, so that, walking in the desert may be convenient. Hoofs are reduced to nail like structure.
4. Limbs are quite long and they are two–toed. There is reduction in the digits. The metacarpals in the fore–limbs

and metatarsals in the hind limbs are fused to form a single bone in each foot known as the *Cannon bone*.

5. Limbs have become secondarily digitigrade.
6. The wall of the stomach has three chambers which serve as water reservoirs (Paunch) convenient for storing and supplying water when needed.
7. Comparatively, there is reduction in the number of teeth. The dental formula of camel is —

$$I\frac{1}{3},\ C\frac{1}{1},\ Pm\frac{3}{3},\ M\frac{3}{3} = \text{Total } 36$$

8. The place of upper two incisors is occupied by horny pad and the lower incisors opposing this pad are spoonshaped.

Place of Origin and Evolutionary Changes

The evolution of camels initiated during the upper Eocene period. North America is undisputedly considered as the evolutionary home of camels because the fossils were mostly obtained from North America. Now, it is well established that during the Pliocene period they invaded South America, different parts of Asia and Africa where they are still surviving but, unfortunately, they became extinct in North America, the place of their origin.

During the course of evolution, several changes took place which have been described below :

1. There was gradual increase in size.
2. Loss of lateral digits took place but finally, there was no sign of vestiges.
3. Elongation and fusion of metapodials took place and as a result cannon bone was formed.
4. Digitigrade foot developed from unguligrade foot.
5. Broadening of Phalanges and development of foot pads took place as adaptation for life in deserts.
6. Reduction took place in the number of teeth and their elongation for grazing purpose has also been marked.
7. Evolutionary changes took place in the skull and feet (as shown in the figures).

Fossil records with their main characters

1. *Protylopus*: This fossil genus was obtained from Eocene period in America. It was a small creature about the size of a jack–rabbit. Hind limbs were longer but the fore–limbs were shorter. The lateral toes of the foot were complete but slender. Ulna and radius were separate. Fibula was complete. There were four functional digits in the hand. Lateral toes of foot were slender. Canines were slightly large and the molars were low crowned. There was altogether 44 teeth.

2. *Poebrotherium*: Its place of discovery was South Dakota. Fossils were obtained from the Oligocene period. It was of the size of the sheep. The limbs were comparatively long. The skull was small and tapering. It possessed a long neck. Limbs were long and the digital reduction was obvious. Coalescence of ulna with the radius was clearly marked. Fibula also coalesced with tibia except at two ends. Some small nodules were present as vestiges of lateral foot. Hoofs resembled with that of deer. Incisors and canines were mammal like and not spatulate, the total number of teeth being 44. Grinding teeth of the upper jaw were short crowned. There was elongation tendency of the molar teeth.

3. *Protomeryx*: The eye orbit was completely encircled by bone and rest of the characters resembled *Poebrotherium*. They were also discovered during the oligocene period. Teeth were suitable for feeding on abrasive grasses.

4. *Paratylopus*: They were considered as the first ancestor of giraffe camel in oligocene period.

5. *Procamelus*: They appeared in Miocene and were larger than the modern Llama. Metapodials were in the process of fusion for forming a connon bone. The first and second incisors were lost in the adults.

6. *Stenomylus* (Gazelle Camel): They were discovered from Western Nebraska during the Miocene period. The head was small and the neck was long and delicate. There were two toes on each foot and the metapodials were not fused. On the lower jaw 10 incisor like teeth were present and out of which 6 were true incisors. Incisor like canines and premolars were also present. They possessed two

toes in each foot. In Pliocene, they evolved into *Rakomylus* which became extinct soon.

7. *Oxydactylus*: The were also known as Giraffe camel. They represented another side branch of camel. They were discovered from the eastern Wyoming and Western Nebraska during the Miocene period. It was smaller in size. Metapodials were not fused to form a cannon bone. Hoofs were short, pointed as in the case of deer. The number of teeth was 44 and they were adapted for browsing. The grinders were short crowned.

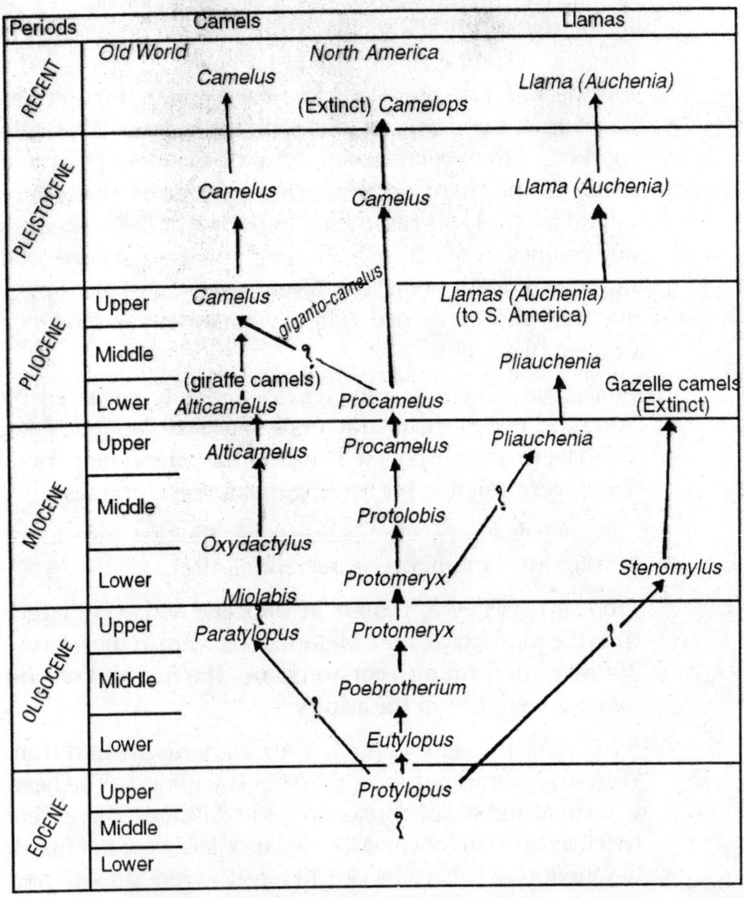

Fig.62 : Evolutionary tree of camel as deduced from available fossils

8. *Alticamelus*: They were also discovered during the Miocene period. They were of larger size in comparison to *Oxydactylus*. They possessed giraffe like long neck and the limbs suitable for desert life. Teeth were low crowned.
9. *Camelops and Eschatius*: Their fossils were available in California during the Pliocene and Pleistocene periods. They were the direct descendants of the Miocene *Procamelus*. Fossils were incompletely known. They were not true camelus but they were definitely camel like.
10. *Camelus*: They were discovered during the Pliocene and Pleistocene from Southern Russia, Rumania and Algeria. They differ from the *Procamelus* of the upper Miocene in loss of one premolar in the upper jaw and premolars in the lower jaw. On the basis of several resemblances it has been concluded that the Pliocene camels are considered as the direct descendents of *Procamelus* of the Miocene period. Pleistocene camels are more *Llama* like. *Procamelus* individuals were capable of facing the cold weather of the north and as such, they passed to the old world through Berring Strait as they have been identified in the Siwaliks in India during lower Pliocene. They are *Camelus sivalensis* and *Camelus antiquus*. Indian species have *Llama* like teeth which are not present in the modern camels. North American camels are supposed to be more primitive.

Chapter 20

Evolution of Man

The modern man *Homo sapiens sapiens* is a member of the family *Hominidae* of order–*Primates*. There is a close similarity between apes and man which have probably evolved simultaneously from a common ancestral stock (*Dryopithecus*). The special features of man which have been acquired during the course of evolution are:

1. Large brain, increased cranial capacity and high intelligence
2. Fore head high and rounded
3. Evolution of hands for manipulation and perfection of thumb opposability
4. Bipedal locomotion and terrestrial habit
5. Upright posture and absence of tail
6. Omnivorous feeding habit
7. Chest is broad due to flat sternum
8. Lumbar curve developed
9. Dental arch is a rounded parabola
10. Legs longer than arms
11. Articulated sound (speech)
12. Scanty hairs on the body
13. Diastema (Simian gap) disappears
14. A chin is present
15. The face is orthognathous in contrast to prognathous of apes (i.e. in apes the face is protruded out into muzzel)

16. The eyebrow ridge has become thin and light
17. Teeth small and closely set

During the *Palaeocene* period three offshoots branched off from some unspecialised arboreal insectivorous strew–like ancestral stock. The three offshoots developed into —

1. Tree shrews,
2. Lemurs and tarsiers and
3. Anthropoid ancestors.

The anthropoid ancestors splitted up from its beginning into three major lines. The first line gave rise to *New World monkeys,* the second developed into *Old World monkeys* and the third was destined to develop into apes and man of the recent epoch. The earliest human fossils were recovered from *Pliocene.*

Descent from the trees

The reasons for coming to the ground from the trees were —

i. Assumption of erect progression
ii. To make the hands organs of mind
iii. Hunting in groups for food
iv. Need for clothing during winter season
v. Use of fibre, and freedom from climatic restrictions
vi. For taking omnivorous diet
vii. Development of Cultural activities

Ancestral Pool for Ape and Man

Now it is well established that the man originated from primitive ape–like forms. The most important ape like forms are regarded as the ancestral pool for both ape and man. They have been described below :

1. *Propliopithecus*: The fossils were obtained from Egypt about 35 millions years ago during the oligocene period. It was a small ape of the size of the squirrel monkey. It is represented by the fossil jaws and teeth. It was ape like and had vertical teeth. A primitive form called *Parapithecus* has also been reported during this period which was almost similar but evolution from

Fig.63 : Early Man

Propliopithecus is more convincing. There are two different ancestors of man's oligocene ancestry —

(i) That *Propliopithecus* directly gave rise to *Ramapithecus*.

(ii) That *Propliopithecus* gave rise to *Dryopithecus* which gave rise to apes and man.

2. *Aegyptopithecus*: It was an extraordinary fossil of the oligocene time which was about the size of modern gibbon. It had expanded cranium, longer face, large canines, spatulate incisors, two premolars and three low crowned molars with five cusps. But it was not considered suitable in the line of evolution towards man.

3. *Dryopithecus*: They were the early hominids present in Asia and Europe during the Miocene period about 30 million years ago. They comprise a group of fossils. It is considered as a common ancestor of man and apes. From this point, the evolution of modern apes and modern man took place in different lines. They had semi erect posture. Complete lower jaw of *Dryopithecus* has been obtained from Bilaspur (H.P.). *Sivapithecus* is another fossil ape from Siwalik hills. They were about the size of modern chimpanzee. Jaws and teeth were less specialised.

4. *Oreopithecus*: They lived in late Miocene or early pliocene period. Fossils were discovered from northern Italy. They resembled more with man. They might have walked erect. It was an aberrant branch from the common ancestors of man and ape. Canines were small in this case.

5. *Ramapithecus* (*Kengapithecus* from Africa): *Ramapithecus punjabicus* was discovered by Lewis from Siwalik Hills of India in 1932. Nature of teeth and studies on jaw fragments revealed that they were the earliest man like primate and they were on direct line of human evolution during the Miocene time. *Dryopithecus* is supposed to give rise to *Ramapithecus*.

Evolution of Man in Pleistocene

(Fossil and Modern Men)

There are three main stages in the evolution of man and they are represented by *Australopithecus* in early *Pleistocene, Homo erectus* in Middle Pleistocene and *Homo sapiens* in the late Pleistocene.

Fig.64 : Gorilla and Chimpanzee

They have been described below :

1. *Australopithecus*: It is also considered as Prehuman phase. They represent the line of human ancestry at the end of Pliocene or at the beginning of Pleistocene. It was discovered in South Africa in 1924 and was called *Australopithecus* by Raymond Dart. The fossils have also been collected from Java. Teeth resembled those of man but they were larger in size. Eye brow ridges projected forward. Cranial capacity was 450-550 cc. They walked upright. The chin was absent. They lived in caves and hunted some animals.

Australopithecines belong to either *gracile* type (e.g. *Australopithecus africanus*), or *robust* type (e.g. *Panthropus*). *Zijonthropus* (from – East Africa), also known as *Australopithecus boisei* and *Meganthropus* (from – Java) are also considered *robust* type.

Homo habilis considered as early human phase was discovered in Tanzania in the year 1964 was almost like *australopithecus* but the cranial capacity was 660 cc. Dentition resembled the modern man. They could manufacture simple tools.

2. *Homo erectus*: Australopithecines were succeeded by *Pithecanthropus*, which were forms with large brains. They were considered as late human phase. Fossils were

Australopithecus

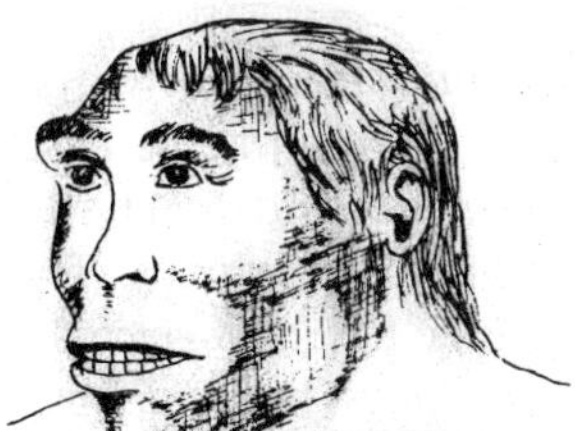

Java man

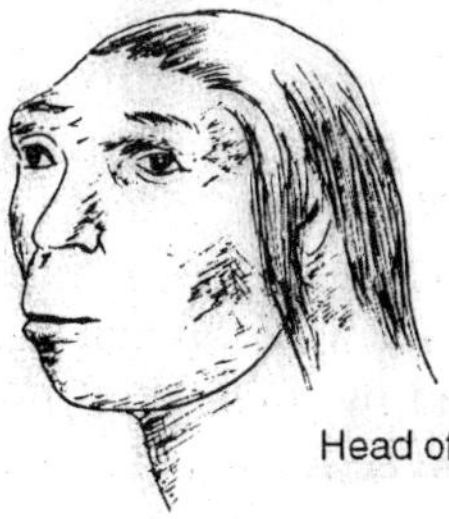

Head of *Peking man*

Fig.65 : Heads of primitive man

obtained from Java by Dubois (1891) during the middle Pleistocene. These were called *Pithecanthropus erectus* (i.e. erect ape–man). Similar fossils were found near Peking (China). They were named *Sinanthropus pekineneis*. Because of similarities with Java–ape–man they were renamed as *Pithecanthropus pekinensis*. These names were however replaced by Mayer (1950) as *Homo erectus*. Thus *Homo erectus*, is called *Homo erectus erectus* and *Pithecanthropus pekinensis* as *Homo erectus pekinensis*.

(i) Java Man (*P. erectus*): Its cranial capacity was 940 cc. It was more than 5 feet in height and was nearly 70 kg. in weight. The forehead was low and slanting. The face was *Prognathous* and the jaws were massive having huge teeth. There was no chin and the bony eye brow ridges were present.

(ii) Peking man (*P. pekinensis*): The fossils were discovered by Davidson Black (1920). Its cranial capacity was 850-1200 cc. Rest of the characters were like Java man. They lived 500000 years ago. They used fire.

3. *Homo sapiens* (late Pleistocene Man): It was the modern human phase. *Homo erectus* gave rise to early *Homo sapiens* which were called *Homo neanderthalensis, Homo heidelbergensis* etc. But now they are grouped under *Homo sapiens*.

(i) *Heidelberg man*: They were discovered by Otto Schoetensack (1908). In this case the lower jaw is massive. It was found in Heidelberg (Germany). Teeth resembled modern man. They are supposed to be contemporary to *Homo erectus* and they gave rise to *Neanderthal man*. The *Heidelberg man* had a cranial capacity of 1300 cc. They knew control of fire and they hunted in groups.

(ii) *Neanderthal man*: Fossils were obtained from Neanderthal Valley in Germany. Previously it was known as *H. sapiens neanderthalensis*. They were present some 150000 years ago. They flourished in Europe, Asia and North America. They were characterised by thick skull bone, low and slanting forehead and heavy eye–brow ridges. There was no chin. The cranial capacity was 1300-1600 cc. They

Restoration of Head of *Neanderthal man*

Cro–Magnon Man

Fig.66 : Heads of primitive man

belonged to modern human phase. They are supposed to have sense of customs and religious rites (particularly at the time of burial of dead bodies).

(iii) *Soloman* (*Homo solonensis*): Its fossils were obtained from the area of Solo river in Java. It might evolved from *Pithecanthropus*. They were more primitive than *Neanderthal man*. They were characterised by heavy ridges over the eyes, and receding forehead. The cranial capacity was 1300 cc.

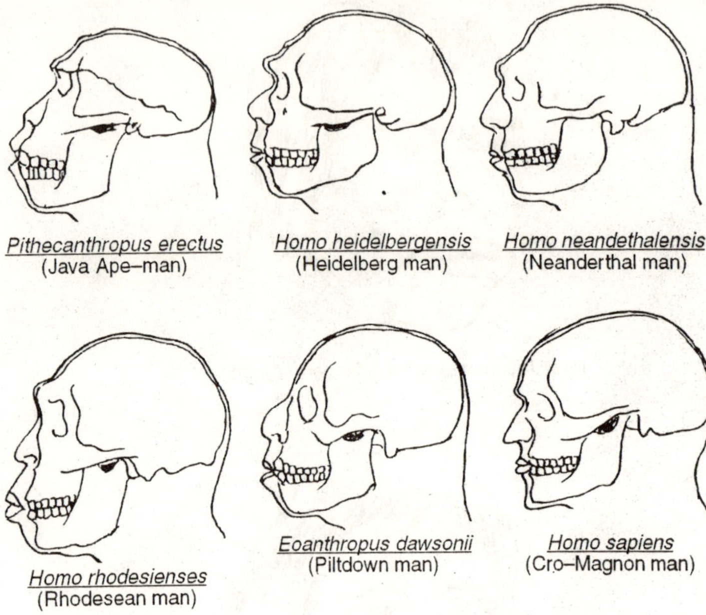

Fig.67A : Reconstructed heads of six human fossils (Redrawn from Organic evolution – Lull)

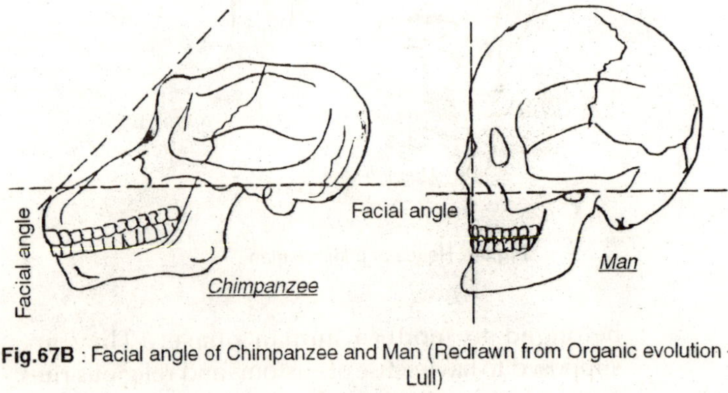

Fig.67B : Facial angle of Chimpanzee and Man (Redrawn from Organic evolution – Lull)

(iv) *Rhodesian man* (*Homo rhodensis*): The fossils were obtained from Rhodesia. The cranial capacity was about 1300 cc. They possessed receding forehead and heavy eye brow ridges.

(v) *Cro–Magnon man*: During the upper Pleistocene, nearly 30000 years ago *Homo sapiens* close to modern

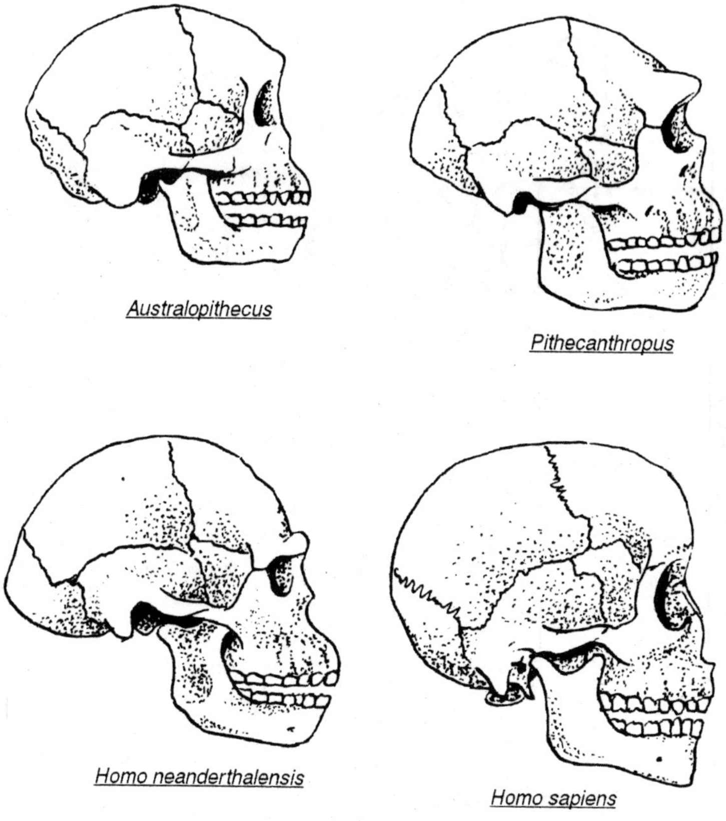

Fig.68 : Skulls of early and modern men

man lived in the cro–Magnon valley in France and other parts of the world. They became extinct about 20000 years ago. They were more advanced than the *Neanderthal man* and belonged to *Homo sapiens*. They had prominent chin. They possessed broad and short face with rounded forehead and narrow nose. The cranial capacity was 1600 cc. They were cave dwelling forms and were characterised by swift locomotion. *Homo sapiens fossils* made sculptures and paintings.

(vi) *Modern man* (*Homo sapiens sapiens*): Further evolution of man after cro–Magnon form is cultural evolution

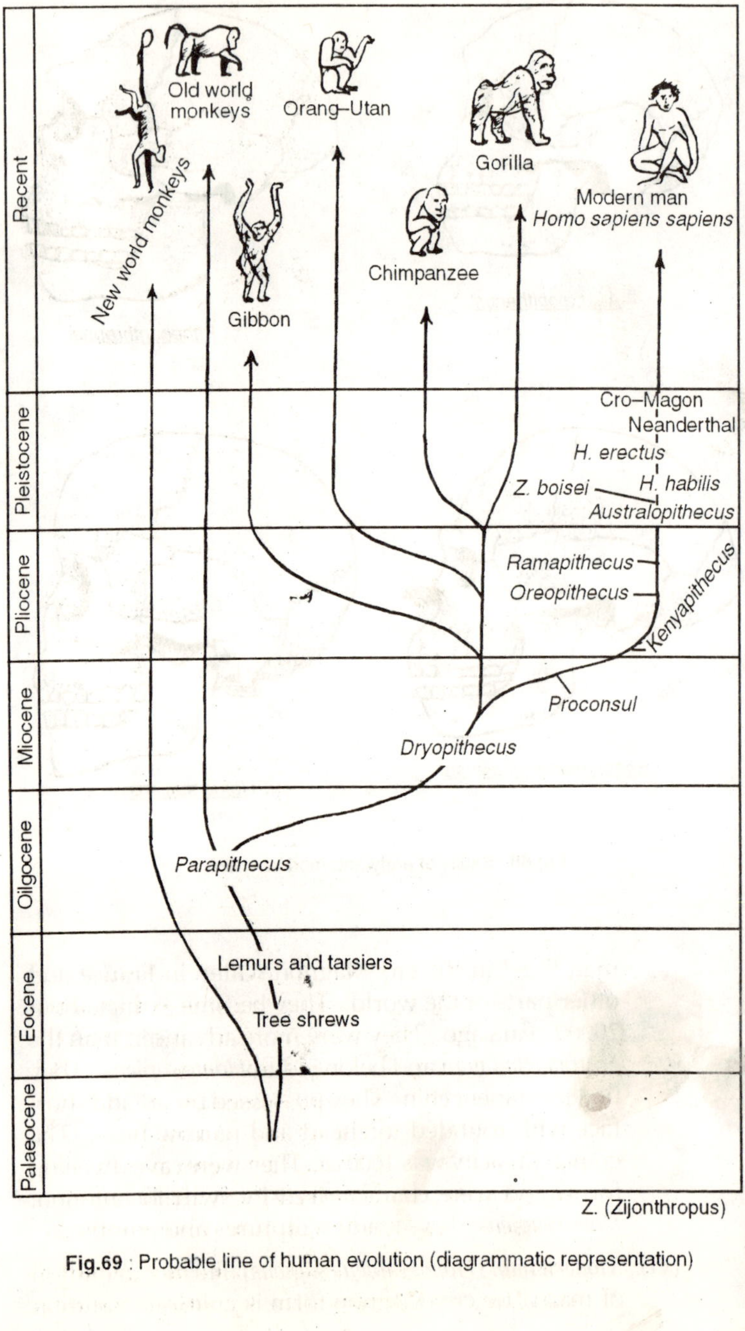

Fig.69 : Probable line of human evolution (diagrammatic representation)

rather than anatomical advancement. The cranial capacity is 1450 ml. They are the true neolithic man. There are five stages of cultural evolution —

(a) *Palaeolithic* – Age of tools of stones and bones – popularly known as stone age.

(b) *Mesolithic* – Age of animal husbandry, language, reading and writing.

(c) *Neolithic* – Age of agriculture, use of clothes.

(d) *Bronze* – Age of use of bronze.

(e) *Iron* – Present age, extensive use of iron.

Chapter 21

Micro–, Macro– and Mega–Evolution

Evolution is essentially a slow and gradual change in the genetic composition of populations. Micro–, Macro–, and Mega–evolutions are three levels of evolutionary process which may also be called "the stages of evolution" or even "the types of evolution". According to Goldschmidt (1940), there is a fundamental distinction between these stages. It is a very crucial point to note that in a continuous population, discontinuity appears which finally leads to marked population differentiation. Isolating mechanisms and some other factors initiate discontinuity within a potentially or really continuous population. Evolution involves minor changes in the gene pool of a particular population from one generation to the other. Modifications in genotypic frequencies, and phenotypic variations become obvious later on. The descending population is not genetically identical to its predecessors. The study of mechanisms of evolution falls within the discipline of population genetics. Changes observed in the subsequent generations of a population may be of different orders of magnitude. Now the different levels of organic evolution have been described below in detail.

Micro–evolution

Micro–evolution indicates relatively small changes in the population. These small changes appear as a result of the interaction of some elemental forces of evolution like mutation, recombination, natural selection and genetic drift. Micro–evolution is a continuous and gradual change in an interbreeding population. Though it is trifling in scale but continuous and gradual changes accumulate to form large differences in structure. The changes are directional for longer periods. Micro–evolution can bring about sub–specific diversification within the species but new species or higher categories of animals can not be produced. This can produce only very small evolutionary differences. Sevage (1965) of the University of Southern California divided the basic patterns of evolution into Sequential and Divergent evolution. The author has considered micro–evolution as Sequential evolution. It is an evolution of modified gene pool from the pre–existing ones. This type of evolution can not produce new population but temporary changes can be brought about. Changes in gene combinations are non–adaptive.

There is change in the gene frequency because of genetic recombination. Gene mutation is either spontaneous or it may be induced with the help of mutagenic agents and environmental factors. Variations can be introduced in the genotype of the individuals by the process of interbreeding where recombination of genes take place. Thus, variations produced in a gene pool by mutation and recombination are operated by natural selection. A number of examples of micro–evolution have been reported in the invertebrate fossils where changes have been observed in the subsequent generations (e.g. in certain species of echinoides and brachiopods). Micro–evolution is the evolution of subspecies or geographic races.

Macro–evolution

Macro–evolution is also known as *adaptive radiation.* Here, divisions of a successful group of animals into smaller groups in response to different habit, habitat and environmental changes take place. The differentiation takes place by adaptive radiation. It has been considered as divergent evolution by Sevage (1969) because the macroevolution results in evolutionary divergence. Specialization takes place in a particular direction. Macromutation takes place in the populations which acquires a new adaptive zone. New population originates from the old ones. The macro–evolution takes much more time in comparison to the sequential pattern. It involves the rise and divergence of discontinuous groups. Whether macro–evolution differs in kind or only in degree from micro–evolution is still controversial. It is definitely a rise of the taxonomic groups having minimum level of genetic discontinuity i.e. upto species and genera levels. In macro–evolution, production of new adaptive types takes place through a process of population fragmentation and genetic divergence.

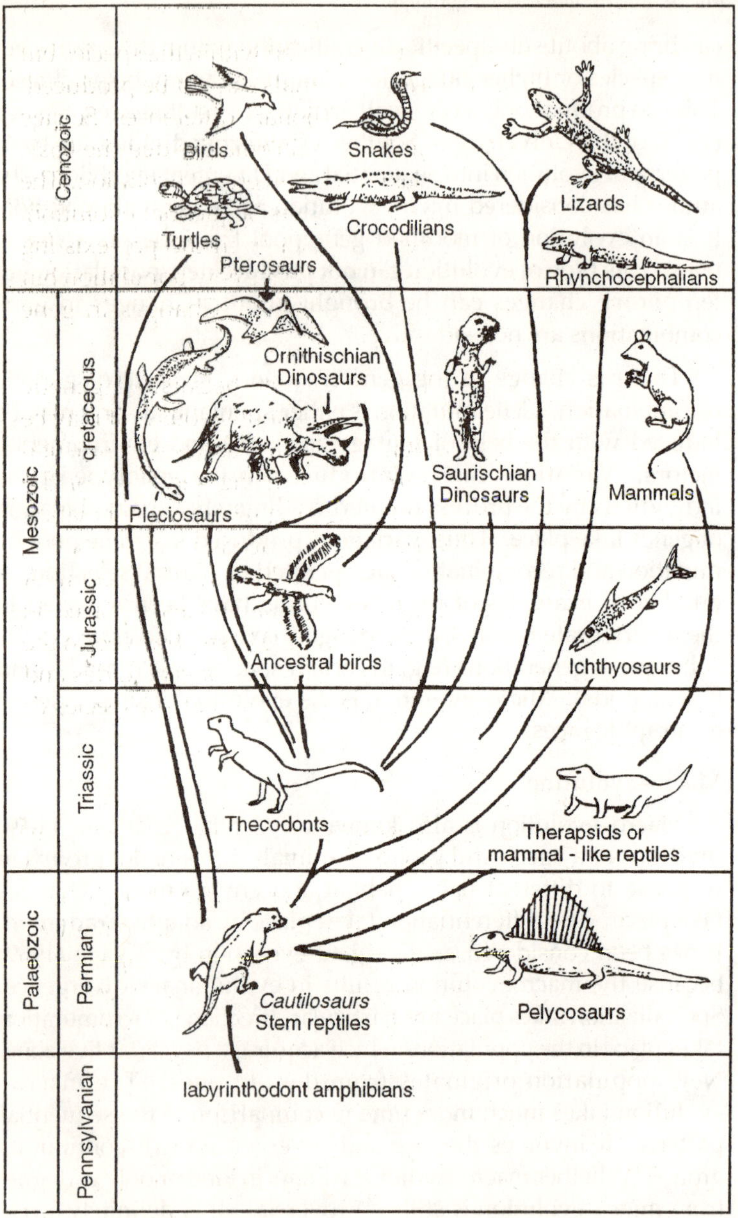

Fig.70 : Adaptive radiation or Macroevolution in reptiles

In macro–evolution, large changes take place in the body form and there is marked increase in the size of the individual also. Organisms become well adapted to changed environments. During the period of geological changes only those organisms which also change, could get a selective advantage and were able to survive in nature. If there are some pre–adaptive characters, they are very useful for the organisms during the transition period. We know, that there were some fishes during the Devonian period which developed lungs and they were proved to be of great use not only in the stagnant drying ponds – but on the land also. Fishes were forced to leave one "adaptive orbit" in order to enter another adaptive orbit. Simpson called it "Quantum evolution" as it was a case of sudden change. Some zoologists have also preferred term "Macroadaptive evolution".

Various organs of the animals change and become differently adapted. Most of these changes are allometric changes, new structures also appear. Thus, changes in body size become obvious after a longer period. Macro–evolution requires sufficient time on geological scales and as such, it is difficult to understand the actual mechanism of macro–evolutionary changes. In macro–evolution, evolution of species and genera takes place.

The best example of macro–evolution will be the evolution of horse. As it is evident from the palaeontological records, the modern horses (*Equus equus*) evolved from the genus *Eohippus* which was not more than 11 inches in height. Number of changes took place in the structure of limbs, length of the neck, length of the neck and structure of teeth etc. Thus, new characters also develop in Macro–evolution. In the divergent evolution several modifications are superimposed upon the basic framework of living material and the spectrum of organism is produced to cope with the changed environment.

Mega–evolution

Megaevolution is the evolution of new types of animals from the predecessors. That means, it gives rise to higher taxonomic categories like orders and classes etc. Now, it is well established that amphibians, originated from fishes, reptiles originated from amphibians and the reptiles gave rise to birds and mammals. These are the examples of Megaevolution. Fishes, amphibians, reptiles, birds and mammals belong to different classes i.e. *Pisces, Amphibia, Reptilia, Aves* and *Mammalia* respectively. So, megaevolution gives rise to new system of animal organization during adaptive radiation. A continuous series of changes take place during the megaevolution. The organisms of the ancestral stock attempt to enter a new zone where there is no competition and also practically there is no such organisms previously belonging to the same group. These organisms entering the new zone develop certain modifications in different directions and finally become suitable for this newly discovered zone for their survival. The organisms of the parental stock develop certain *preadaptational features* which enable them in entering this new adaptive zone. From there they may radiate into several habitats suitable for their survival. Development of more specialised adaptations are known as postadaptations. The characteristic features of Megaevolution can be summarised as given below on the ideas of Sevage (1969) :

1. That the experimentation and exploration of new zone by the members of ancestral stock takes place in several divergent lines. New characters appear to cope with the demands of the new zone.
2. Megaevolution operates on individuals which have developed some suitable preadaptations for the new zone.
3. These individuals then cross the ecological barrier and makes a breakthrough into the new zone rapidly in order to avoid extreme negative selection.
4. The initial shift is always followed by adaptive radiation which is nothing but macroevolution.

The evolution of mammals (class – Mammalia) can be traced from a series of fossil *Synapsid* reptiles of the group therapsida. During this process of evolution, the structure of many parts of the body changed in mammalian direction. False palate, differentiation of teeth into incisors, canines, premolars and molars, and modification of limbs so as to bring the elbows and knees under the body were the important mammalian characters. These changes appeared in the reptilian forms. But, the teeth were without roots and the quadrate and articular bones did not take part in the formation of the ear–ossicles of the middle ear.

These characters are peculiar to mammals only. The quadrate and the articular bones became free from the original function and this was because of the reason that the dentary was elongated and the jaw articulation was taken over by the squamosal bone. Thus, the free quadrate and articulars facilitated the formation of a chain of bones in the middle ear which is considered as a case of pre–adaptation. The formation of ear–ossicles was an adaptation for improving hearing. Articulation of mandible or dentary with the skull could strengthen the jaw. Thus large changes in Megaevolution are due to adaptational needs of the organisms.

Concluding Remarks

Most of the changes are non–adaptive at the level of microevolutions. As the populations are continuously undergoing adaptive radiation, most of the changes are because of the macro– and megaevolutions. Adaptive characters arise as a result of environmental changes. When the animals enter into a new adaptive zone, they must possess such characters as to make them capable of living under changed conditions. Whether a particular animal can adjust in a new environment or not, is determined by the characters it possesses. The suitability of the organism for some changes in habits and habitats are actually 'Preadaptations'. No adaptation is possible without some preadaptations.

According to Goldschmidt, Preadaptation is due to macro–mutations which enable the organisms to switch on to new life. The another view is proposed by others is that the origin is by Post–adaptation in the original environment. But it is a fact, that higher categories come into being through megaevolutionary changes mostly. Micro–evolutionary changes can be understood in terms of the known genetic agents but the macroevolutionary changes are limited to genetic levels. Some authors are of opinion that microevolution and macroevolution are relative terms having only descriptive meanings.

All the three evolutionary processes differ from one another to a considerable degree but they all are based on microevolutionary process. Though all contribute to adaptations, however, macro– and mega–evolutions have some additional forces involving adaptive radiation and divergence.

Chapter 22

Mutations

According to Hugo de Vries, *mutation* meant a striking change of phenotype which was heritable. It was considered as an inception of the hereditary variations. But the term *mutation* is now used to cover only those changes which alter the chemical structure of the gene at the molecular level. Mutation may occur suddenly in the populations reproducing in natural way. Mutation is thus a hereditary change in the genetic composition of the individual, other than which may be caused by the simple recombination of genes. There are two types of mutations – in one type there are changes in the structure and composition of gene, known as *gene mutations* or *point mutations* and in another type there are changes in the structure or number of chromosomes, known as *chromosomal mutations*. Sometimes, it becomes difficult to distinguish between gene mutation and chromosomal mutation, as the phenotypic effects may be similar in both the cases. Mutation can affect all kinds of traits – external or internal, morphological, physiological or biochemical. It plays an important role in the origin of new species.

Mutation Theory of Hugo de Vries

This theory was based upon the observations on evening primose, *Oenothera lamarckiana*. It was observed that the wild variety of primose showed characteristic differences and they could be classified into eight categories which were regarded as separate species. They are —

1. *O. gigas*,
2. *O. rubrinervis*,
3. *O. nariella*,

4. *O. brevistylis,*
5. *O. laevifolia,*
6. *O. oblonga,*
7. *O. albida,* and
8. *O. lata.*

All these species had distinct morphological differences. Hugo de Vries noted that these characters originated suddenly and were inheritable. The new characters were seen in subsequent generations. The new species formed were callea *mutant* and the newly developed characters were called *mutations.* On the study of these forms the *mutation theory* was formulated by him. According to Hugo de Vries, new species originate as a result of large discontinuous variations which appear suddenly and form new species at once. Salient features of this theory are —

1. Mutations arise suddenly from time to time in the naturally breeding populations and the mutants are different from their parents.
2. Mutations are heritable and as a result of mutations, new species are formed.
3. Mutations are not directional.
4. Mutations are large and quite different from Darwin's small and directional fluctuating variations.
5. Mutations are always subjected to Natural Selection.
6. Unsuitable mutants do not survive in nature, when acted upon by Natural Selection.
7. Mutations may be dominant or recessive, and the latter remain hidden for several generations.

Criticism of Mutation Theory

That the sudden appearance of species differences as proposed by Hugo de Vries was criticized by Darwinians who believe that evolution results from gradual fluctuating inheritable differences over a series of generations. Davis claimed that *Oenothera lamarckiana* was of hybrid origin which produced several varieties. T.H.Morgan has stated that in plants like *Oenothera* the origin of species by mutation is not rare case rather this happens in many plants, but it does not occur in animals. Morgan (1909) showed that mutations could be of any magnitude in Drosophila. Hugo de Vries did not lay stress on genic mutations.

In addition to these facts, mutations can not explain properly the process of adaptations. It is difficult to believe that each individual has appeared suddenly. Presence of flightless birds on oceanic islands can not be explained with the help of mutations. Today mutations are noticed among all organisms from viruses to man. Thus mutation was exposed to severe criticism. But now, it has been realised that mutations are significant to evolutionary process. Mutation alone can not account for evolution. It contributes much in the origin of new species but only when the other forces in nature act simultaneously.

Details of Classification of Mutations

There are different types of mutational changes which are

1. *Gene mutations or Point mutations*: In this case there is a change in the chemical organisation of gene. Change in the base sequence of DNA takes place. There can be an *addition* or *deletion* of nucleotides and *substitution* of nucleotides.

2. *Chromosomal mutations*: It involves losses, multiplications or spatial arrangement of genes in the chromosomes (*Chromosomal aberrations*).

   (a) *Inversion*: It is the intrachromosomal segmental rearrangement.

   (b) *Translocation*: It is the interchromosomal segment rearrangement.

   (c) *Duplication*: Duplication of chromosomal segment

   (d) *Delition*: It involves the loss of a segment of chromosome.

3. *Change in the number of chromosomes (Heteroploidy)*:

   (a) *Haploidy*: Decrease in the number of chromosomes sets (n – chromosome).

   (b) *Polyploidy*: Increase in the number of chromosomes (Triploidy – 3n, Tetraploidy – 4n, Pentaploidy – 5n).

4. *Somatic Mutations*: Huskins has illustrated certain cases of somatic mutations resulting in patches of tissues. But,

usually, the word mutation is taken to mean a genetic mutation.

5. *Spontaneous Mutations*: That, some mutations arise spontaneously in animals and plants. It is a natural spontaneous event. There are two important features of this type of mutation :

   (a) Repeated occurrence of the same mutation

   (b) Reversibility of the process of mutation

6. *Induced Mutations*: These mutations are produced by the induction of certain *mutagenic agents*. There is experimental control over the process of mutation and the process of mutation can be induced at will in a particular population. The following are the mutagenic agents —

   (a) *X–rays*: Muller (1927) demonstrated that X–rays can induce mutations.

   (b) *Ultraviolet rays*: Unlike X–rays, ultra–violet rays penetrate tissues only very slightly. Pyrimidines are very sensitive to ultraviolet rays in comparison to purines.

   (c) *Temperature*: Mutation rate is increased by heat, but, some mutations increase with the low temperature also.

   (d) *Chemical agents*: Some chemical agents are known to be mutagenic e.g. Mustard gas, β–propiolactones, diazomethane etc.

   (e) *Irradiated media*: Some investigators at the University of Texas observed that bacteria and neurospora grown in ultraviolet irradiated media showed increase in mutation rate.

Modern Concept of Mutation / Gene Mutation

The gene is composed of DNA. So, 'Mutation' is a change in the gene or DNA. The DNA molecule has a coded information in the form of a particular sequence as base pairs. A change in the sequence of arrangement by pyrimidines and purines changes the function of the gene, causing change in the phenotype. Watson and Crick have suggested that mistakes in pairing might occur at the time of duplication of DNA. Mistake in pairing leads to mutation and may occur by one of the following methods :

1. *Base analogue*: In certain cases, the analogue can take the place of the natural substances in the same biological reactions. For example, 5 *bromouracil* (5BU) is an analogue of *Thymine* and is mutagenic in bacteriophage and bacteria. If *Escherischia coli* is deprived of thymine and grown in the presence of BU, the latter is incorporated in to the DNA and under suitable conditions, the thymines can be replaced with its analogue as shown below :

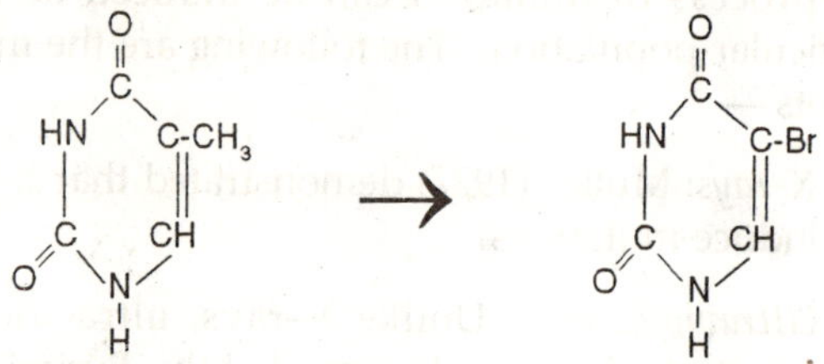

Thymine *5–Bromouracil*

Suppose that the sequence of bases within a gene is

A – T
G – C
– A – T
and, C – G

A mutation will be a change in the sequence resulting in

A – T
G – C
– G – C
C – G

Hence, the third nucleotide pair has been changed from A–T to G–C. In analogue like 5–BU can bring about this change in 3 steps at the point of DNA duplication. Thus, G pairs with C. This change from A–T to G–C is the result of an error on the second step when the analogue BU pairs with G instead of A.

The following is the graphical representation of mutation after *Levine* (1962).

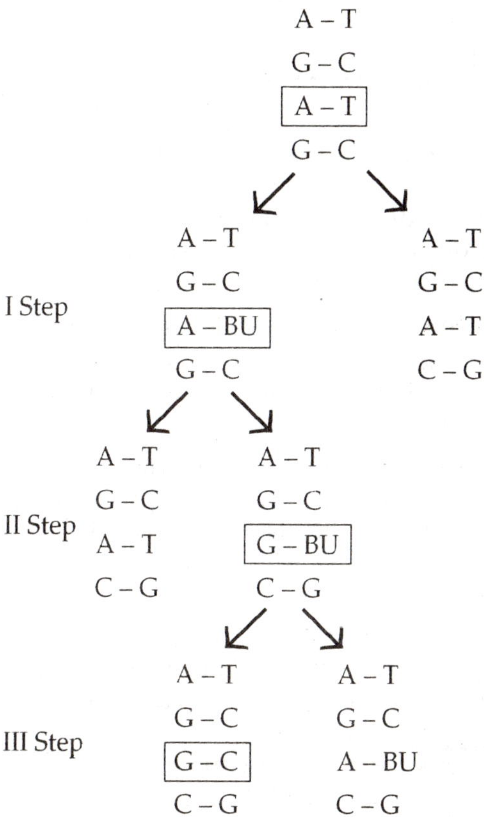

2. Modification of Purines and Pyrimidines:

i) *By Nitrous oxide* (HNO_2): — Some Chemicals act as mutagen and change the bases of DNA molecule. HNO_2 is a very good example. It acts as mutagen by oxidative

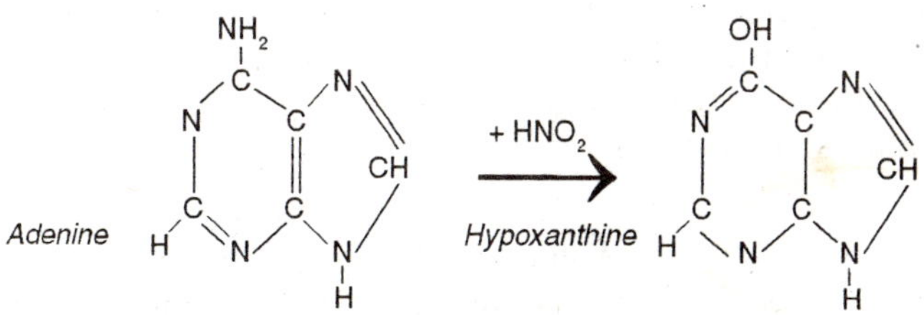

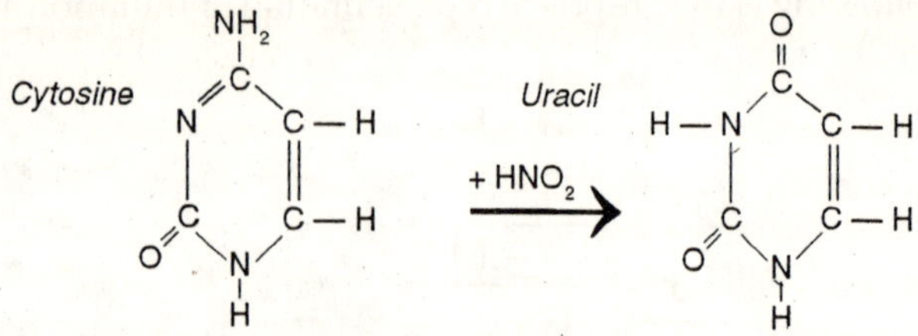

deamination of hydroxyl (OH) group for an amino (NH_2) group in the organic base. For example, *adenine* can be changed to *hypoxanthine* which causes mutation.

Thus a new sequence of bases would be carried over at the time of duplication. The pairing behaviour of hypoxanthine is like that of *guanine*. Hypoxanthine pairs with *Cytosine* rather than with thymine.

Similarly, Cytosine can be deaminated to form *Uracil* which has pairing properties. The hydrogen bonding properties of Uracil are similar to those of thymine. So, there is U–A pairing. Also, guanine is deaminated to *Xanthine* which behaves like guanine and pairs with Cytosine.

Instead of G–C pairing there is X–C pairing. Deamination of guanine is not mutagenic. The following table shows the pairing behaviour of DNA following the deamination of nitrous oxide.

Normal bases of DNA	*Normal pairing*	*Bases formed by deamination*	*New pairing*
Adenine	A–T	Hypoxanthine	G–C
Cytosine	C–G	Uracil	U–A
Guanine	G–C	Xanthine	X–C

Adenine and cytosine have a different pairing behaviour. As a result, changes in the DNA take place in 50% of the progeny. Deamination of guanine does not result in a heritable mutation because there is no change in the pairing behaviour of the deaminated base (xanthine).

ii) *By hydroxylamine* (NH_2OH): NH_2OH reacts with cylosine and guanine residues and brings about *transitions* and mispairing. During transitions a purine is replaced by another purine and a pyrimidine is replaced by another pyrimidine. Deamination of cytosine to a base takes place and it pairs with adenine instead of guanine. Thus, C–G pairing is changed into A–T pairing.

iii) *By Alkylating agents*: Some mutagenic agents, for example — *diethyl sulphate, methyl methane sulphate* and *ethyl ethane sulphate* etc. are included among these agents. The alkylation reaction takes place at the N–7 position of guanine residues or at the N–3 position of adenine residues. Pairing error takes place. Hydrolysis takes place in the base–pair in the base–sugar linkage and as a result, it releases the base from the DNA molecule. This creates gap in one chain.

3. *Distortions in DNA*: Mutation may be caused by deletion or addition of bases by certain *acridine* dyes such as proflavin and acridine orange. The acridines are flat molecules, like purine bases and can be intercalated between the bases of the DNA helix. This distorts the structure of DNA and can result in deletion or insertions of bases during recombination. Here the new DNA has a additional base pair. As a result of the acridine molecule between two bases of the template strand is formed in the lengthening of the DNA molecule. The addition or deletion of nitrogenous bases in the DNA or mRNA is known as *frame–shift mutations* because they shift the reading frame of codons from the site of change onward. Thus, deletions or insertions of nucleotides take place in DNA molecule at one or more places which cause mutations.

4. *Radiation as a physical mutagen*: It is a well established fact that high–energy radiation can change the atomic structure of a substance. It causes loss of an electron and the formation of an *ion*. Alterations in nucleic acids may be caused by radiation which are of great genetic importance. Ionizing radiation has greater penetration power in comparison to UV radiation and produces free radicals which tend to mobilize molecules. Nucleic acid absorb UV–light, causing their nitrogen — containing bases to become highly reactive free radicals. UV–radiation causes addition of water molecule to pyrimidines in both DNA

and RNA forming photophosphates. X–rays bring about mutations by breaking the phosphate ester linkage in DNA. As a result, a large number of bases are lost (deletion) or rearranged.

Chapter 23

Zoogeographical Regions

The distribution of animals, both invertebrates and vertebrates on the surface of the earth is very peculiar. Most of the naturalists including Scalter (1857), have divided the terrestrial portion of the earth's surface into six primary zoological regions which corresponds with the continental masses of geographers. Wallace (1876) accepted Scalter's scheme of geographical distribution of animals with slight modifications. It is practically impossible to study the invertebrates and all the animals inhabiting the sea. Only limited study is possible. There are six zoogeographical regions (realms) which have been described below with their boundaries.

1. *Palaearctic region*: This region embraces Europe, northern corner of Africa, a part of north Asia and numerous oceanic islands of the North Atlantic. This region has been divided into four sub–regions i.e. *European, Mediterranean, Siberian* and *Manchurian*.

2. *Ethiopian region*: This region comprises whole of Africa (South of tropic of cancer), the southern portion of Arabia, Madagascar and some adjacent islands. It consists of four sub–regions — i.e. *East African, West African, South African* and *Malagsy*.

3. *Australian region*: This region embraces the continent of Australia, Tasmania, New Zealand, New Guinea, Eastern island of Indonesia (Hawaii islands) and some islands of the Pacific. This region consists of four sub–regions, i.e. *Australian, Austro–Malayan, Polynesian* and *New Zealand*.

4. *Oriental region*: This region comprises India, Burma, Ceylon, Siam, South–east China, Thailand and Malayan islands (Jawa, Sumatra, Borneo, Philippines etc.). It

consists of five sub–regions i.e. *Indian, Ceylonese, Indo–Chinese, Malayan* and *Indo–Malayan.*

5. *Nearctic region*: This region embraces greater part of North America, Greenland, Iceland etc. It consists of four sub–regions i.e. *Californian, Rocky mountain, Canadian* and *Alleghany.*
6. *Neotropical region*: It comprises South America with Central America, greater part of Mexico, and West Indies. The sub–regions include — *Chilian sub–region, Brazilian sub–region, Mexican sub–region* and *Antilean* (West Indies) *sub–region.*

Fauna of Different Regions

Palaearctic Region

Fishes: *Acerina cornua* (small perch), *Precarina, Aspro* (zingels) — all belonging to family percidae are present. *Brachymystax* (Icconus) family Salmonidae is very common. *Carassius* (Gold fish), *Tinca* (Tenches), *Rhodeus* (Bitterlings), *Misgurnus* and *Noemacheilus* Sp. are also found. Paddle fish *Polydon* occurs in China and *Dallia* (black fish) is found in eastern Siberia. *Esox lucius* (Northern pike) is also very commonly found in this region.

Thus several species of cobitids, anabantids, cyprinids, catfishes, channids, Mastacembelids, and Percids have been reported. Sticklebacks, Sturgeons and Pikes are also present.

Amphibians: Among the amphibians are — Giant Salamander (*Megalobatrachus*), Fire Salamander (*Salamendra salamandra*), *Proteus, Hynobius* (Hynobiidae), Mud puppy (*Necturus*), *Siren, Amphiuma, Cryptobranchus, Newts, Bombina, Discoglossus,* Midwife food (*Alytes*), *Rana, Rhacophorus, Bufo, Pelodytes* (Pelodytidae).

Reptiles: Among the snakes — *Coronella austriaca* (smooth snake), *Vipera berus* (European viper), True viper, Asian Pit Vipers, *Typhlops* and *Sand Boas* are very common.

Among the lizards, the Land lizard, *Lacerta agilis*, and Green lizard, *Lacerta viridis* are very common.

The *Trionyx* of eastern asia, *Alligator* of China, *Chameleon* and Varanus are also found in this region.

Birds: The important bird fauna of this region includes — European Night Jar, *Caprimulgus europaeus;* European Avocet, *Recurvirostra avoseta,* Hedge–sparrows, *Accentor;* Beared Tit,

Panurus; European Jays, *Garrulus glandarius*; European Green Finch, *Carduelis chloris*; *Fringilla coelebs* (Fringillidae), European oyster catcher, *Haematopus ostralagus*; Partridges, *Perdix*; Pheasant, *Phasianus* (Phasianidae); Golden pheasant, *Chrysolophus*; Common Quail, *Coturnix coturnix*; European Eagle owl, *Bubo bubo*; and Hoopoe, *Upupa epops*.

Ducks, Cuckoos, Kingfisher, Woodpecker, Thrushes, Crows, Starling, Warblers, Waxwings, Vulture, Hawks, Herons, Loons, Larks etc. are also found in this region.

Mammals: Among the mammals are included — European Hamster, *Cricetus cricetus*; Yak, *Poephagus*; Seal, *Phoca sibirica*; Desman, *Myogale*; *Elephant shrews*, Civet, Hyaena, Chamois, *Rupicapra rupicapra*; *Camelus*, *Talpa* (mole), Serotine Bat, *Eptesicus serotinus*; Long eared bat, *Plecotus auritus*; *antelopes*, *Mustela*, Great Panda, *Ailuropus*; Tibetan Langur, *Rhinopithecus*; Tufted deer, *Elaphodus*; chinese water–deer, *Hydropoteus*; Crested Porcupine, *Hystrix cristata*; Roe deer, *Capreolus capreolus*; Common Dormouse, *Muscardinus avellanarius*; Edible Dormouse; *Glis glis*; European mole rat, *Spalax microphthalmus*; Reindeer, *Rangifer tarandus* etc.

Ethiopian Region

Fishes: Several cat fishes, cyprinodonts and Chichlids are present in this region. African lungfish, *Protopterus* (Protopteridae) is a rare fish of this region. African electric fishes (Mormyridae), *Notopterus* (Notopteridae) and *Calamoichthys* (Reedfish) are also found. Fishes of the family Gymnarchidae are also on record.

Amphibians: *Xenopus* (Pipidae) is peculiar to this region. Caecilians are also present. Members of the family Hylidae are absent. Nectophrynoides are also present.

Reptiles: Important among the snakes are Puff Adder (*Bitis arientans*), *Typhlops punctatus*, Tree snakes (*Dispholidus typus*), *Boaedon dineatum*, *Dsypeltis scabra*, *Licheterodon madagascariensis*, *Python sebae*, *Python regius*, *Naja haje*, *Naja nigricolis* and several snakes of family elapidae, Mambas (*Dendroaspis*) etc. Some more snakes of the families Colubridae and Elapidae are also found.

Lizards are represented by African horned Chameleon — *Chamaeleo oweni*, Sungazer — *Cordylus giganteus*, *Agama* — *Uromastyx acanthinurus*, *Agama stellio*, Fat tailed Gecko — *Hemitheconyx caudicinctus*, skink — *Trachysaurus rugosus*, *Uroplatus fimbriatus*, etc.

Important crocodiles are — *Crocodylus niloticus* and *Osteolaemus tetraspis*.

Kinixys erosa — fresh water hinge–back Tortoise of the family Emydidae and *Chemmys leprosa* are the characteristics of this region.

Birds: There are several birds peculiar to this region. Some important among them are —

Plantain (*Musa*) eater (Musophagidae), Mouse birds — Colies (Collidae), *Leptosomus* (Leptosomatidae), *Pitta angolensis* (Pittidae), painted stork — *Ibis leucocephalus*, *Philepitta* (Philapittidae), *Bufagus* (sturnidae), Shoebill — *Balaeniceps rex*, Spoon bill — *Platalea alba*, Long–legged Secretary bird — *Sagittarius serpentarius*, *Pycnonotus* sp. (Picnonotidae), Flycatchers (Muscicapidae), Vulturine Guinea fowl — *Acryllium Vulturinum*, Bee–eaters — *Meropus apiaster* (Meropidae), Pinnant–winged Night jar — *Semeiophorus* (Caprimulgidae), and Long–tailed Nightjar — *Scotornis climacurus* (Caprimulgidae). Besides them, Cackoos (cuculidae), crows (Corvidae), vultures, Eagles, Ostriches (Struthioniformes), Hornbills, Hoopoes, Buffalo weavers, Hammer head birds, Irrisor and some other birds are also found.

Mammals: The important mammals of this region are — Aard vark, *Orycteropus* (Orycteropodidae); *Hippopotamus* (Hippopotamidae), Giraffe (Giraffidae), Hyraxes or Dassies (*Procavia capensis*), African flying squirrel, *Anomalurus* (Anomaluridae); Aardwolf, *Proteles* (Protelidae); crested rat, *Lophiomys*; Aye–eye, *Daubentonia madagascarinsis*; several genera of *Civets*, otter shrews, *Potamogale velox* (Potamogalidae), Moles, *Chrysochloris* sp. (Chrysochloridae); Brazza's Monkey, *Cercopithecus neglectus*; Thumbless monkey, *Colobus*; *Gorilla*, Chimpanzee, Baboons, *Theropithecus neglectus*; small–scaled Pangolin, *Manis tricuspis*; Jumping hare, *Pedetes cafer*; Greater Leaf–nosed bat, *Hipposiderus commertoni*; Ring–tailed Lemur, *Lemur catta*; Zebra, *Equus burchelli*; Arabian Camel, *Camelus dromedarius*; African white Rhinoceros, *Diceros simus* (Rhinocerotidae); African elephant, *Loxodonta africana*; Servaline cat, *Leptailurus serval*; Desert Lynx, *Caracal caracal*; Deer, Goat, Sheep (All–cervidae) etc.

Australian Region

Fishes: Important fishes are — *Neoceratodus* (Dipnoi), Glove eels, *Alabes* (Synbranchidae); *Arapaima* sp. and *Heterotis* sp. (Both belonging to Osteoglossidae).

Besides these fishes, members of the families Atherinidae, Trachinidae, Gadopsidae, Siluridae and Haplochitomidae etc. are also present.

Amphibians: Important amphibians of this region are —

Liopelma (Liopelmidae), *Crinia* (Leptodactylidae), *Myobatrachus gouldii*, and *Hemisus marmoratus*. Rana and Toads are basically absent, however, they have been secondarily introduced at certain places. Amphibians are of course very few in number.

Reptiles: Among the snakes, Pythons (Pythonidae) and Cobra (Elapidae) are very common. Crotalidae and Viperidae are absent from this region.

Among the lizards, Australian Frilled lizard — *Chlamydosaurus kingi*, Beared lizard — *Amphibolurus barbatus*, Scale footed lizard — (Pygopodidae) and Naked–toed Gecko — *Gymnodactylus miluisi* are very common. Monitors (Varanidae) are also found in this region. Member of families Pygopodidae and Aprasiidae are also on record. *Sphenodon* or tuatara (Rhynchocephalidae) is confined to Newzealand.

Among the tortoise and turtles, Long–necked swamp Tortoise — *Chelodina longicollis*, River turtle — *Carettochelys* (of New Guinea) are very common. *Chelemys* and *Elseya* are also found.

Birds: Important birds of this region are — Paradise birds (Paradisaedae), Pied Fantail —*Rhipidura rufifrons*, Honey Suckers (Meliphagidae) Emu — *Dromaius novaehollandiae*, Scrub birds (Atrichidae), Zebra Finch — *Poephila castanotis* (Fringillidae), Cockatoos — *Kakotoe galerita*, Paraquets (Platycercidae), Kea, *Nestor* (Psittacidae); Brush–tongued parrot (Trichoglossidae), Kiwi — *Apteryx haasti*, Megapodes (Megapodiidae), Rock hopper — *Eudyptes crestatus*, Cassowaries (Casuarridae), Flycatchers — *Rhipidura* and *Pachycephala* (Muscicapinae), Australian Pelicans — *Pelicanus conspicillatus*, Cuckoo–shrike — *Coracina lineata* (Compephagidae), Wood Swallows (Artamidae), The Takahe — *Notornis mantelli*, Flower peckers (Dicaedae), Pigeons (Columbidae), Rednecked Avocet — *Recurvirostra novaehollandiae*, King fishers (Alcaedinidae), crested Pigeon — *Ocyphaps lophotes*, Honey suckers (Meliphagidae), Newzealand Wren or Rifleman — *Acanthisitta chloris*, etc.

Mammals: The important mammals of this region are — Tree Kangaroo — *Dendrolagus lumholtzi*, Marsupial mole

(Notoryctidae), whiptail small Kangaroo — *Wallabia elegans*, Wambat — *Vombatus hirsutus* (Vambatidae), striped Phalangers — *Dictylopsida picata* (Phalangeridae), Long–nosed bandicoot – *Perameles nausta* (Peramelidae), Duck bill — *Ornithorhynchus anatinus* (Ornithorhynchidae), Echidna — *Tachyglossus aculeatus* (Tachyglossidae), Apes and Monkeys, Tiger cat — *Dasyurops macculatus*, Tasmania Devil — *Sarcophilus harrisii*, Rusty Numbat — *Myrmecobius rufus* etc.

Oriental Region

Fishes: The important fishes which belong to this region are Catfishes (Siluridae), Carps (Cyprinidae), *Mastacembelus* (Mastacembelidae), *Channa* Sp. (Channidae), *Nandus nandus* (Nandidae), Air-breathing fishes, Eels (Anguilliformes), Perciformes, Cobitidae, Tetradontiformes, Syngnathidae, Notopterydae, Belonidae, etc.

Amphibians: Important amphibians are Caecilians (Caecilidae), Salamanders (Salamandridae), Discoglossidae, Pelobetids, Toads and True frogs etc.

Reptiles: Characteristic snakes are – Cobra, Krait (Both Elapidae), Typhlopes, Pythons (Pythonidae), Pit vipers (Crotalidae), Whip snake – *Dryophis prasinus* (Dryophidae); Tree snakes (Dendrophidae), Nocturnal tree snakes (Dipsadidae) etc.

Other important reptiles are – Geckos (Gekkonidae), Chameleon (Chamelendiae), Varanus (Varanidae), Agamids (Agamidae), Skinks (Scincidae), Crocodiles, *Crocodilus porosus* (Crocodilidae), Gavialis, *Gavialis gangeticus* (Gavialidae), Freshwater tortoise and turtles (Emylidae) etc.

Birds: Important among the birds are – common Myna – *Acridontherus tristis* (Sturnidae), Black Myna – *A. ginginianus* (Sturnidae), Pied Myna–*Sturnus Contra* (Sturnidae), House Crow – *Corvus splendens* (Corvidae), Jungle crow – *C. macrorhynchus*, (Corvidae), Bulbul – *Picnonotus* spp. (Picnotidae), Blue Peacock – *Pavo cristatus*, Indian cuckoo–*Eudynamys scolopaeca* (Cuculidae), Blue rock Pigeon – *Columba livia* (Columbidae), Horned Owl – *Bubo bubo* (Strigidae), Himalayan Barred Owlet – *Glaucidium cuculoides* (Strigidae), Egret – *Egretta garzetta*, Pond Heron – *Ardeola grayii* (Ardeidae), Honey eater (Meliphagidae), Grey leg goose – *Anser anser* (Anatidae), Bar headed goose – *Anser indicus* (Anatidae), Small blue king fisher – *Alcedo attahis* (Alcedinidae), Green Bee–eater – *Meropus orientalis* (Alcedinidae), Hoopoe –

Upupa epops (Upupidae), Adjutant stork – *Lepoptios dubius* (Ciconidae), Lesser adjutant – *L. Javanicus* (Ciconidae), Scavanger vulture – *Neophron percnopterus* (Accipitridae), Griffon vulture – *Gyps himalayensis* (Accipitridae), Brahminy kite – *Haliaster indus* (Accipitridae), Rose–ringed parakeet – *Psittacula krameri* (Psittacidae), Red kite – *Milvus milvus* (Accipitridae), Shikra – *Accipiter badius* (Accipitridae), Flower pecker – *Dicaeum erythorhynchus* (Dicaeidae), Woodpecker – *Dryobates himalayensis* (Picidae), House sparrow – *Passer domesticus* (Ploceidae), Black Drongo – *Dicrurus adsimilis* (Dicrutidae), Jungle Babbler – *Turdoides striatus* (Muscicapidae), Common Warbler – *Prinia subflava* (Muscicapidae), Munia – *Estrilda* sp., Jungle fowl. – *Gallus*, etc.

Mammals: The characteristic mammals of oriental region are – Indian elephants – *Elephas indicus, Elephas maximus*, Slow loris – *Nycticebus concang*, Tarsier – *Tarsius spectrum*, Apes (Orangutan and gibbons), Desert Lanyx – *Caracal caracal*, Flying lemurs – *Galeopithecus* (Galeopithecidae), Tree Shrews (Tupaiidae), Apes – (Orang–utan–*Simia Satyrus*, Gibbon – Hylobatidae), Tiger – *Felis tigris*, Malayan tapir – *Tapirus indicus*, Sloth bear – *Melurus, Horned* antilope – *Tetraceros*, Mongoose – *Herpestes edwardsi*, Lion – *Felis leo*, Domestic Cat – *Felis domestica*, Fishing Cat – *Felis viverrina*, Indian desert cat – *Felis lybica ornata*, Leopard Cat – *Felis bengalensis*, Langurs – *Presbytis*, Rhesus monkey – *Macaca mulatta*, Rhinoceroses (Rhinocerotidae), Buffaloes, horses (*equus*), Bats, dogs, pigs, camels, (*camelus*) etc. are also very common.

Nearctic Region

Fishes: Important fishes are – *Polydon* (Paddle fish), Lepidosteus (garpike), Brown Bull head – *Ameiurus nebulosus*, Dwarf Perch of California – *Micrometrus minimus*, Bowfin – *Amia*, Large mouth Bass – *Micropterus selmoides, Paralabrax* (Perch), *Pomotis* (Perch), *Baleosoma, Noturus* (Siluridae), *Thaleichthys* (Salmonidae), Strugeons etc.

Amphibians: Among the amphibians are – *Amphiuma, Bufo amenicanus, Siren*, Spotted Salamander – *Ambystoma maculatum, Axolotl*, Frogs (Ranidae), *Hyla*, also, there are more than nine genera of the family Salamandridae.

Reptiles: Among the snakes, Ribbon snake – *Thampnophis sauritus*, (Colubrid snake), Banded Raffler – *Crotalus horridus* (Crotalidae), Copper head – *Ancistrodon contortrix* are the characteristic of snakes of this region.- *Chilomeniscus, Pituophis* and *Conophis* etc. are also present. Lizards are represented by Geckos, *Ophiosaurus, Phrynosoma Euphryne, Uta, Heloderma, Uma* etc.

Among Tortoises, *Aromochelys* and *Chelydra* are very common. The yellow mud turtle or Terrapin – *Kinosternon flavescens* is the characteristic reptile fauna. Red eared turtle – *Pseudemys cripta* is also found in this region. Alligator snapping turtle – *Macrochelys temmincki*, American Alligator – *Alligator mississipiensis*, and American Crocodile – *Crocodylus acutus* are also the characteristic reptiles.

Birds: The important birds of this region are – Black and White warbler – *Mniotilta varia* (Parulidae), Humming birds (Trochilidae) – Ruby – Throated Humming bird – *Archilochus colubris*, Rufous Humming bird – *Selaphorus rufus*; Woodpeckers (Picidae), Yellow–bellied sapsucker – *Sphyrapicus varius*, Red–headed woodpecker – *Melanerpes erythrocephalus*, Golden Fronted woodpecker – *M. aurifrons*, Northern three–toes woodpecker – *Picoides tridactylus; Sialia sialis* (Paridae), Loons – *Gavia immer*, Ruffed Grouse – *Bonasa umbellus, Chamaea fascitata* (Muscicapidae), Red cross bill – *Loxia curvirostra* (Fringillidae), North American Killdeer – *Charadrius vociferus* (Charadriidae).

Besides the above mentioned characteristic birds, Pelicans, Hawks, Vulture, Herons, Pigeons, Ducks, Quails, Owls, Kingfisher, Waxwings, Wrens, and Turkeys are also very common.

Mammals: The important mammals of the Nearctic region are – Virginian Opossum – *Didelphys virginiana*, Armadillo – *Dasypus novemcinctus* (Dasypodidae), Pronghorn –*Antilocapra americana* (Antilocapridae), North American Bison – *Bison bison* (Bovidae), Musk–ox – *Ovibos moschatus* (Bovidae), Pack rats – *Neotoma* Spp. (Muridae), Flying squirrel – *Glaucomys sabrinus* and *G. volans* (Sciuridae), Prairie Dog – *Cynomys ludovicianus* (Sciuridae), Ground hog – *Marmota monax* (Sciuridae), Chimpmunk – *Tamias striatus* (Sciuridae), Porcupine – *Erethizone dorsatum* (Erethizontidae), American badger – *Taxidea taxus* (Mustelidae), American mole – *Scalapus aquaticus* (Chrysochloridae), American beaver – *Castor canadensis* (Castoridae), Grizzy bear – *Ursus horribilis* (Ursidae), American Elk (Wapati) – *Cervus canadensis* (Cervidae), American Moose (*Alces americana*), Red Bat – *Lasiurus borealis*, Starnosed Mole – *Condylura cristata*, Meadow Mouse – *Microtus pennsylvanicus*, white footed Mouse – *Peromyscus leucopus* and Shrews – *Some araneus*. Besides these, Rocky mountain goat (*Haplocercus*), *Haplodon* (Haplodontidae), Arctic fox and Polar bear are also present in this region.

Neotropical Region

Fishes: Among the fishes, Electric eels – *Electrophorus electricus, Gymnotus* – (Gymnotidae), Rays (Trygonidae) and Silurids (Siluridae) are very common. Carp family (Cyprinidae) is completely absent. *Lepidosiren paradoxa*, Garpike – *Lepidosteus* and characian fish are also present.

Amphibia: Among the amphibians, *Siphonopsis* (Caecilidae) *Rhinotrema* (Caecilidae), *Speleurpes* (Urodele), Tree Frogs (Hylaplesidae), Surinam toad (Pipidae), *Rana* (Ranidae), *Rhinophrynus* (Rhinophrynidae), and *Axolotl (Ambystoma mexicanum)* are important.

Reptiles: *Typhlops*, Boa constrictor – *Constrictor constrictor, Boa canina*, Pit vipers (Crotalidae), Coral snakes Elops (Elapidae), Lyptotyphlops are the important snakes of this region.

Among the lizards, sungazer – *Cordylus giganteus*, Geckos and Iguans – *Iguana tiguana*, Heloderma (Helodermatidae), chirocolidae, Cercosauridae, Iphisidae, etc. are common. Among the crocodilia and chelonia, few species of *Crocodylus* and Alligators and chelonians are present. *Sternotherus odoratus* (Kinosternidae), *Hydromedusa, Chelys fimbriata* (Chelidae), *Podocnemis expansa* (Pleomedusidae), Black Caiman – *Melanosuchus niger* (Alligatoridae) are very common.

Birds: The important birds are – Black–necked swan – *Signus melanocoryphus, Tinamous crypturellus varigenatus, Endromia elegans*, South American Ostrich – *Rhea Americana*, Momot (Momotidae), Todies (Todidae), Sparrow–sized plant cutters (Phytotomidae), Chatterers – *Rupicola* (Contingidae), Puff–bird – *Notharchus macrorhynchus* (Bucconidae), Jacanas – *Jacana spinosa* (Jacanidae) Oil–bird – *Steatornis caripensis* (Steatornithidae), Curassows – *Crax rubras* (Cracidae), Hoazin – *Opisthocomus hoazin* (Ophistocomidae), Trumpeters – *Psophia crepitans* (Psophiidae) and Sun bitterns – *Eurypyga helios* (Eurypygidae). Some common birds like Ducks, Herons, Plover, Storks, Woodpeckers, Swallows, Kingfishers, Pigeons and Herons are also there.

Mammals: There are several members belonging to the family *Cebidae*. They are white collared Titi – *Callicebus torquatus*, Masked

Titi – *C. personatus*, White headed Saki – *Pithecia*, Black headed Uakari – *Cacajao melanocephalus*, Red Uakari – *C. rubicundus*, Bald Uakari – C. *Calvus* and several species of Howling monkeys (*Alouata* spp.). The Great Blood sucking or Great Vampire Bat – *Desmodus rotundus* and Lesser Blood scuking or Lesser Vampire Bat – *Diphylla ecaudata* (Both family – desmodontidae) are also present. Chinchillas – *Chinchilla laniger* (Chinchillidae), Brazilian cavy – *Cavia operea* (Caviidae), Rodents (Echimyidae), six banded armadillo – *Euphractus sexicantus* (Dasypopidae), Three–toed anteater – *Tamandua tetradactyla* (Myrmeco–phagidae) are also found in this region. Besides the above mentioned mammals, Ring tailed Coati – *Nasua rufa* (Procyonidae), Terrestrial Short-tailed opossum – *Peramyus*, water opossum – *Chironectes minimus* (Didelphidae) and Solenodon – *Solenodon paradoxus* (Solenodontidae) are the characteristic mammalian fauna of this region.

Wallace's Line

Though, six Zoogeographical regions were clearly identified however, it was difficult to draw a line of distinction between the Oriental and Australian regions. The main massses of these two regions are separated from one another by strings of islands collectively known as *Malaya Archipelago*. The islands of Sumatra, Borneo and Java belonged to the Oriental region. They are separated from Malaya and from one another by shallow sea and their animals are very similar to those of Malaya. In the East, New Guinea and the Aru islands are included in the Australian region. These are-separated from one another by shallow sea. There are several animals in common with that of Australia. There are other smaller islands like Philippines, Celebes, the Moluccas, Timor etc. whose exact Zoogeographical areas were to be assigned.

In the middle of the 19th century, Wallace studied the fauna of Malaya, Borneo and Bali islands. In 1856 he travelled from Bali to Lombok located at a distance of twenty miles. Marked difference was observed in the fauna of these two islands. Bali had the fauna of Oriental region and Lombok was reported to have the fauna of Australian region. Later on, the fauna of Celebes, the Kei island, the Aru islands, Timor, the Moluccas and New Guinea were also studied by him. On the basis of faunestic peculiarities, he classified them as being Oriental or Australian.

In 1863 he drew a line on the map which marked the boundary between Oriental and Australian region. This line was popularly known as Wallace's line. This line was drawn on the map between the Philippines and the Moluccas in the North, then South–west between Borneo and Celebes, and South between the small islands of Bali and Lombok.

Weber's Line

Weber drew another line between the Oriental and Australian regions some years after the Wallace's line was drawn. Weber's line was drawn eastwards separating Moluccas from Celebes and Kei island from Timor.

Thus, there are two schools of Zoogeographers, one prefers Wallace's line whereas another school has approved the Weber's line. There is transitional fauna between the two lines, some are Oriental in nature and others are related to Australian region. Wallace's line, indicates the eastern limit of what was a land mass joined to Malaya, whereas Weber's line marks off the western limit of what was originally a part of Australian continent.

It has been suggested that the islands like Celebes, Flores and Lombok constitute the transitional region which is neither the part of Oriental region nor the part of Australian region. They were

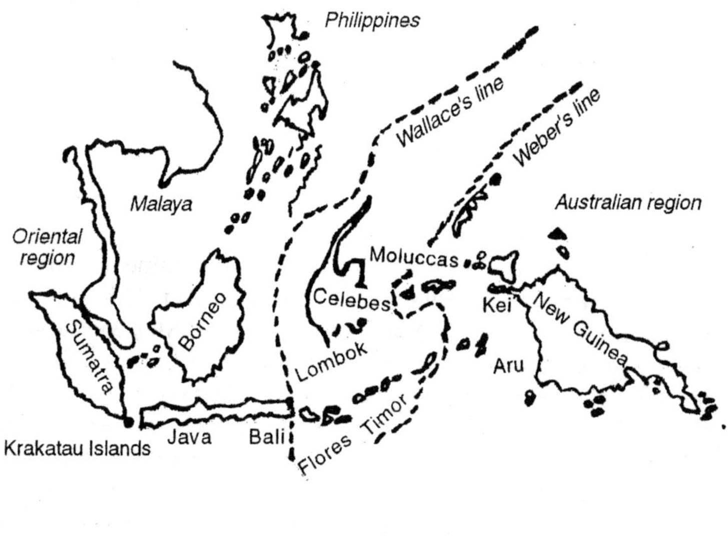

Fig.71 : Malaya Archipelago, showing Wallace's line and Weber's line

probably under the sea for most of the Coenozoic and lost their earlier flora and fauna. They appeared as independent islands. Migrants from Australian and Oriental regions are mostly present in this transitional region. So, they possess their characteristic fauna now. They are considered as separate region called *Wallacea*.

Classification of Islands

The distribution of animals on the islands presents complicated patterns but it is very much interesting at the same time. Islands have been separated into two categories – *Continental islands* and *Oceanic islands*. The continental islands are originally the part of a large land msss or continent but got separated later on from the parent land by a stretch of sea either by the sinking of land or by a rise in the level. The fauna of continental island shows much similarities with the fauna of main land particularly when the island is a young one. Diversified fauna may appear later on when the process of evolution takes palce independently. The important continental islands are – *Great Britain, Formosa, Madagascar, Borneo* and *Japan*. Out of these islands, *Great Britain* is comparatively young.

The *oceanic islands* had never a connection with the land mass or continent. They originated independently. They have a volcanic origin or they were formed by the activity of corals and sometimes by both ways. The oceanic islands differ from the nearest continent in their climate, vegetation and fuana. Even closely located oceanic islands are not identical. Some important oceanic islands are – *Galapagos islands, Azores, Bermuda, St. Helena* and *Sandwich* islands etc.

Continental Islands and their Fauna

1. Great Britain

Great Britain in supposed to be a young continental island. British Isles consists of Britain or U.K. and Ireland. U.K. consists of England, Wales and Scotland. At the end of Miocene, British Isles became separated and formed a part of Europe with no landmass connection (*Britain* is about 89000 sq.miles and *Ireland* covers about 32000 sq.miles). The area is poorly represented by Cricetids (*Cricetus cricetus*), shrews, hedgehoges, moles, squirrels, hares, murid rats and mice, cats, weasels, reindeer, red deer, bats, badgers, otters etc. Avian fauna is represented by sea birds.

Limbless lizard *Anguis* (Anguidae) is also present. Snakes could not reach Ireland area particularly. Red Grouse – *Lagopus scoticus* (Tetraonidae) is the characteristic fauna.

2. New Zealand

New Zealand is situated about 1000 miles south east of Australia. It comprises two main islands and cover an area of about 100000 sq. miles. The funa of New Zealand is small but remarkable. Mammals are mostly absent, only two families of bats are present, one being purely endemic. Among the birds, flightless birds like *Kiwi,* and Rhea are present. *Flightless rails,* and *Owl parrot* are also there. Species of giant Moas became extinct. *Goose* and *Wrens* (both flightless) are also present. Wattlebirds are also common. *Kea* – a large parrot is peculiar to this continental island. Among the reptiles *Geckos* and *Skinks* are found in New Zealand. A living fossil, *Sphenodon punctatum* (tuatara) belonging to order *Rhynchocephalia* is found in this island only. *Liopelma* is peculiar among the amphibians. Freshwater fishes are completely absent.

3. Madagascar

Madagascar is located in the Indian Ocean, about 260 miles from east coast of Africa. It covers an area of approximately 24000 sq. miles. Among the mammals Madagascar has many bats of African and Oriental affinities. Several Insectivores, Primates (*aye–aye*), Rodents and Carnivores are present. About 10 genera of lemurs forming a natural group and not related with other genera found elsewhere are present here. A small *hippopotamus* occurred in Pleistocene which was different from the present day *hippopotamus* of Africa. Some more mammals have been introduced recently (e.g. Shrews, Bush Pig, Murid mice etc.). The avian fauna includes four species of Asities and false sunbirds (Philepittidae). These are – *Philepitta castanea, P. schlegeli, Neodrepanis coruscans* and *N. hypoxantha*. Vanga shriks (Vangidae) are also confined to Madagascar. Giant elephant birds (now exitnct) and the rail like mesoenatids are the flightless birds present in Madagascar. Among the reptiles are present the turtles, geckos, chamaeleons, skinks and spiny lizards. The poisonous snakes are absent. The American family of lizards, Iguanidae is also present in Madagascar. Tree frogs are endemic fauna among the amphibians. Fishes are absent.

Oceanic Islands and their fauna

1. Galapagos Islands

Galapagos islands are oceanic islands of the pacific sea which are about 600 miles west of south America. These are volcanic emerging from deep water on the equator. The islands cover an area of 2870 sq.miles. The fauna is scanty. Freshwater fishes and Amphibians are absent. Reptiles are represented by *giant land tortoises,* large *iguanas,* small *iguanids, colubrid snakes* and *geckos.* The avian fauna comprises the *Darwin's finches* constituting a natural group of several varied forms derived from a single ancestor from South American mainland and radiated in various directions in accordance with their feeding habits. They are –

1. Cactus feeders,
2. Ground teeders,
3. Insectivorous tree finches,
4. Woodpecker – like finches,
5. Vegetarian tree finches.

Accordingly the modifications in the beaks are well marked. Flightless *cormorants* and species of *penguins* are also present on these islands. Among the mammals are present the species of bat and erecetid rat.

St. Helena

St. Helena is a volcanic oceanic island covering an area of 80 sq.miles. It lies in South Atlantic Ocean 1000 miles from Africa. The fauna is very poor. There is complete absence of mammals, reptiles, amphibians and freshwater fishes. The native bird is a *Plover.* Insects and land molluscs are found in abundance.

Index

B

C

D

E

F

I

N

O

P

T

U

V